This book comes with access to more content online.

Test your knowledge with a quiz
for every chapter!

Register your book or ebook at
www.dummies.com/go/getaccess.

Select your product, and then follow the prompts
to validate your purchase.

You'll receive an email with your PIN and instructions.

Anatomy & Physiology Workbook

3rd Edition with Online Practice

by Erin Odya and Pat DuPree

for dummies®

A Wiley Brand

Anatomy & Physiology Workbook For Dummies®, 3rd Edition with Online Practice

Published by: John Wiley & Sons, Inc., 111 River Street, Hoboken, NJ 07030-5774, www.wiley.com

Copyright © 2018 by John Wiley & Sons, Inc., Hoboken, New Jersey

Published simultaneously in Canada

For general information on our other products and services, please contact our Customer Care Department within the U.S. at 877-762-2974, outside the U.S. at 317-572-3993, or fax 317-572-4002. For technical support, please visit https://hub.wiley.com/community/support/dummies.

Wiley publishes in a variety of print and electronic formats and by print-on-demand. Some material included with standard print versions of this book may not be included in e-books or in print-on-demand. If this book refers to media such as a CD or DVD that is not included in the version you purchased, you may download this material at http://booksupport.wiley.com. For more information about Wiley products, visit www.wiley.com.

Library of Congress Control Number: 2018935444

ISBN 978-1-119-47359-6 (pbk); ISBN 978-1-119-47366-4 (ebk); ISBN 978-1-119-47358-9

Manufactured in the United States of America

SKY10025612_031121

Contents at a Glance

Table of Contents

Introduction

Whether your aim is to become a physical therapist or a pharmacist, a doctor or an acupuncturist, a nutritionist or a personal trainer, a registered nurse or a paramedic, a parent or simply a healthy human being — your efforts have to be based on a good understanding of anatomy and physiology. But knowing that the knee bone connects to the thigh bone (or does it?) is just the tip of the iceberg. In *Anatomy & Physiology Workbook For Dummies*, 3rd Edition, you discover intricacies that will leave you agog with wonder. The human body is a miraculous biological machine capable of growing, interacting with the world, and even reproducing despite any number of environmental odds stacked against it. Understanding how the body's interlaced systems accomplish these feats requires a close look at everything from chemistry to structural mechanics.

Early anatomists relied on dissections to study the human body, which is why the Greek word *anatomia* means "to cut up or dissect." Anatomical references have been found in Egypt dating back to 1600 BC, but it was the Greeks — Hippocrates, in particular — who first dissected bodies for medical study around 420 BC. That's why more than two millennia later we still use words based on Greek and Latin roots to identify anatomical structures.

That's also part of the reason so much of the study of anatomy and physiology feels like learning a foreign language. Truth be told, you are working with a foreign language, but it's the language of you and the one body you're ever going to have.

About This Book

This workbook isn't meant to replace a textbook, and it's certainly not meant to replace going to an actual anatomy and physiology class. It is designed as a supplement to your ongoing education and as a study aid in prepping for exams. That's why we give you insight into what your instructor most likely will emphasize as you move from one body system or structure to the next.

Your coursework might cover things in a different order than we've chosen for this book. We encourage you to take full advantage of the table of contents and the index to find the material addressed in your class. Whatever you do, certainly don't feel obligated to go through this workbook in any particular order. However, please do answer the practice questions and check the answers at the end of each chapter because, in addition to answers, we clarify why the right answer is the right answer and why the other answers are incorrect; we also provide you with memory tools and other tips whenever possible.

Within this book, you may note that some web addresses break across two lines of text. If you're reading this book in print and want to visit one of these web pages, simply key in the web address exactly as it's noted in the text, pretending as though the line break doesn't exist. If you're reading this as an e-book, you've got it easy — just click the web address to be taken directly to the web page.

Foolish Assumptions

In writing *Anatomy & Physiology Workbook For Dummies*, 3rd Edition, we had to make some assumptions about you, the reader. If any of the following apply, this book's for you:

» You're an advanced high school student or college student trying to puzzle out anatomy and physiology for the first time.

» You're a student at any level who's returning to the topic after some time away, and you need some refreshing.

» You're facing an anatomy and physiology exam and want a good study tool to ensure that you have a firm grasp of the topic.

Because this is a workbook, we had to limit our exposition of each and every topic so that we could include lots of practice questions to keep you guessing. (Believe us, we could go on forever about this anatomy and physiology stuff!) In leaving out some of the explanation of the topics covered in this book, we assume that you're not just looking to dabble in anatomy and physiology and therefore have access to at least one textbook on the subject.

Icons Used in This Book

Throughout this book, you'll find symbols in the margins that highlight critical ideas and information. Here's what they mean:

TIP

The Tip icon gives you juicy tidbits about how best to remember tricky terms or concepts in anatomy and physiology. It also highlights helpful strategies for fast translation and understanding.

REMEMBER

The Remember icon highlights key material that you should pay extra attention to in order to keep everything straight.

WARNING

This icon — otherwise known as the Warning icon — points out areas and topics where common pitfalls can lead you astray.

EXAMPLE

The Example icon marks questions for you to try your hand at. We give you the answer straightaway to get your juices flowing and your brain warmed up for more practice questions.

Beyond the Book

In addition to the material in the print or e-book you're reading right now, this product also comes with some access-anywhere goodies on the web. While it's important to study each anatomical system in detail, it's also helpful to know how to decipher unfamiliar anatomical terms the first time you see them. Check out the free Cheat Sheet by going to www.dummies.com and typing for "*Anatomy & Physiology Workbook For Dummies* cheat sheet" in the Search box.

You also get access to our online database of questions with even more practice for you. It contains an interactive quiz for each chapter, allowing you to hone your new knowledge even more!

To gain access to the online practice, all you have to do is register. Just follow these simple steps:

1. **Register your book or ebook at Dummies.com to get your PIN. Go to** www.dummies.com/go/getaccess.

2. **Select your product from the dropdown list on that page.**

3. **Follow the prompts to validate your product, and then check your email for a confirmation message that includes your PIN and instructions for logging in.**

If you do not receive this email within two hours, please check your spam folder before contacting us through our Technical Support website at http://support.wiley.com or by phone at 877-762-2974.

Now you're ready to go! You can come back to the practice material as often as you want — simply log on with the username and password you created during your initial login. No need to enter the access code a second time.

Your registration is good for one year from the day you activate your PIN.

Where to Go from Here

If you purchased this book and you're already partway through an anatomy and physiology class, check the table of contents and zoom ahead to whichever segment your instructor is covering currently. When you have a few spare minutes, review the chapters that address topics your class already has covered. It's an excellent way to prep for a midterm or final exam.

If you haven't yet started an anatomy and physiology class, you have the freedom to start wherever you like (although we suggest that you begin with Chapter 1) and proceed onward and upward through the glorious machine that is the human body!

1

The Building Blocks of the Body

Learn the language of anatomy and physiology.

Explore the basic building blocks and functions that make the parts of the body what they are. Dig into atoms, elements, chemical reactions, and metabolism.

Crack open the cell to see what's happening at life's most fundamental level. Find out about the cell membrane, the nucleus, organelles, proteins, and the cell life cycle.

Plunge into cell division, which has several phases: interphase, prophase, metaphase, anaphase, telophase, and cytokinesis.

Use histology to build all of the body's tissues — epithelial, connective, muscular, and nervous — from the inside out.

Chapter **1**

The Language of Anatomy & Physiology

Human *anatomy* is the study of our bodies' structures while *physiology* is how they work. It makes sense, then, to learn the two in tandem. But before we can dive in to the body systems and their intricate structures, you must first learn to speak the language of the science.

Organization of the Body

As you know, the body is organized into systems, grouping together the organs that work together to achieve a common goal. To house all these organs, our body must create spaces to hold them. The body has two cavities that achieve this: the *dorsal cavity*, which holds the brain and spinal cord and the *ventral cavity* that holds everything else. The dorsal cavity splits into the *spinal cavity*, which holds the spinal cord, and the *cranial cavity* that houses the brain. The ventral cavity is split into the *thoracic cavity* and the *abdominopelvic cavity* by a large band of muscle called the *diaphragm*. Within the thoracic cavity are the right and left *pleural cavities*, which hold each lung, and the *mediastinum*. Within the mediastinum is the *pericardial cavity* which contains the heart. The abdominopelvic cavity divides into the *abdominal cavity* (with the stomach, liver, and intestines) and the *pelvic cavity* (with the bladder and reproductive organs), though there's no distinct barrier between the two.

In order to create these cavities within our bodies, we have membranes to border the space. The *visceral membrane* lies atop of the organs, making direct contact with them. For example, the outermost layer of the heart is called the *visceral pericardium* and on the lungs it's the *visceral pleura*. The *parietal membrane* lies on the other side of the spaces or lining the cavity itself. So the lining of the abdominopelvic cavity is known as the *parietal peritoneum* (note that it's not the parietal abdominopelvic that just sounds weird).

The other parts of the body are divided into *axial* and *appendicular* areas. The axial portions are the parts of your body that form your axis — the head, chest, and abdomen. The appendicular portions form your appendages — your arms and legs. For consistency when referencing them, there are proper terms for all of the body's areas. The terminology used in identifying many of the regions is found in Table 1-1. You'll notice these terms popping up all over this book.

Table 1-1   The Body's Regions

Proper Term	Region	Proper Term	Region
Antebrachial	forearm	Genicular	knee
Antecubital	inner elbow	Inguinal	groin/inner thigh
Axillary	armpit	Lumbar	lower back
Brachial	upper arm	Mental	chin
Bucchal	cheek	Orbital	eye
Carpal	wrist	Otic	ear
Cephalic	head	Pectoral	chest
Cervical	neck	Pedal	foot
Coxal	hip	Plantar	sole/bottom of foot
Crural	shin	Popliteal	back of knee
Cubital	elbow	Sural	calf
Dorsum	back	Tarsal	ankle
Femoral	thigh	Vertebral	backbone
Frontal	forehead		

That's a lot of new terms for the first chapter! Let's see how well they're sticking.

EXAMPLE

Q. Which of the following organs would you find in the mediastinum?

    I. lungs

    II. heart

    III. liver

    a. I only

    b. II only

    c. III only

    d. I & II

    e. I, II, & III

A. The correct answer is only the heart. The mediastinum is defined as the area between the lungs and the liver is in the abdomino-pelvic cavity.

1-10 Label the body cavities illustrated in Figure 1-1.

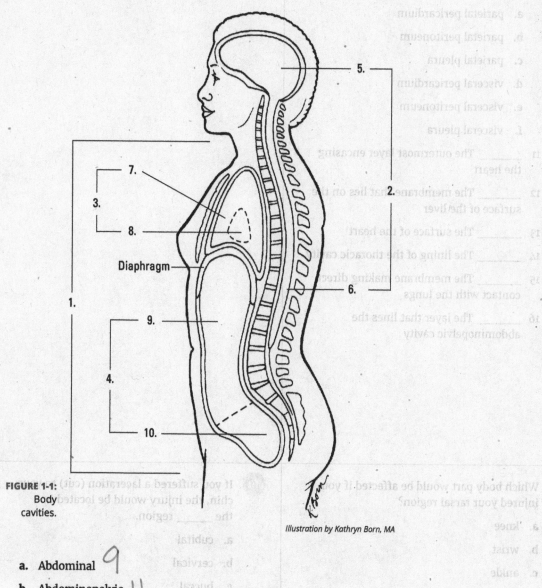

FIGURE 1-1:
Body
cavities.

Illustration by Kathryn Born, MA

a. Abdominal 9
b. Abdominopelvic 4
c. Cranial 5
d. Dorsal 2
e. Pelvic 10
f. Pericardial 8
g. Pleural 7
h. Spinal 6
i. Thoracic 3
j. Ventral 1

11–16 Match the description to identify the membranes that create the body's cavities.

a. parietal pericardium

b. parietal peritoneum

c. parietal pleura

d. visceral pericardium

e. visceral peritoneum

f. visceral pleura

11 ___A___ The outermost layer encasing the heart

12 ___E___ The membrane that lies on the surface of the liver

13 ___R___ The surface of the heart

14 ___C___ The lining of the thoracic cavity

15 ___F___ The membrane making direct contact with the lungs

16 ___B___ The layer that lines the abdominopelvic cavity

17 True or False: The cephalic region is considered part of the appendicular body.

18 Which body part would be affected if you injured your tarsal region?

a. knee

b. wrist

c. ankle

d. shoulder

e. hip

19 If you suffered a laceration (cut) to your chin, the injury would be located in the _____ region.

a. cubital

b. cervical

c. buccal

d. mental

e. frontal

20 Identify the correct pairing of terms:

a. popliteal – inner elbow

b. lumbar – back of the neck

c. antecubital – upper arm

d. coxal - shoulder

e. sural – back of lower leg

Getting into Position

In anatomy and physiology, we often identify the body's features in reference to other body parts. Because of this, we need a standardized point of reference, which is known as *anatomical position*.

REMEMBER

Anatomical position is the body facing forward, feet pointed straight ahead, arms resting on the sides, with the palms turned outward. Unless you are told otherwise, this is the body's position whenever specific body parts are described in reference to other locations.

Because we can only see the external surface of the body, sections must be made in order for us to see what's inside. It's important to take note of what type of section was made to provide the view you see in a picture or diagram. There are three planes (directions) in which sections can be made:

» **frontal:** separating the front from the back

» **sagittal:** dividing right and left sides

» **transverse:** creating top and bottom pieces

We also use directional terms to describe the location of structures. It helps to learn them as their opposing pairs to minimize confusion. The most commonly used terms are:

» **anterior/posterior:** in front of/behind

» **superior/inferior:** above/below

» **medial/lateral:** closer to/further from the midline (also used with rotation)

» **superficial/deep:** closer to/further from the body surface

» **proximal/distal:** closer to/further from attachment point (used for appendages)

WARNING

Right and left are also used quite often but be careful! They refer to the patient's right and left, not yours.

You got it? Let's find out.

21-23 Identify the planes of body sections in Figure 1-2.

    a. Sagittal

    b. Transverse

    c. Frontal

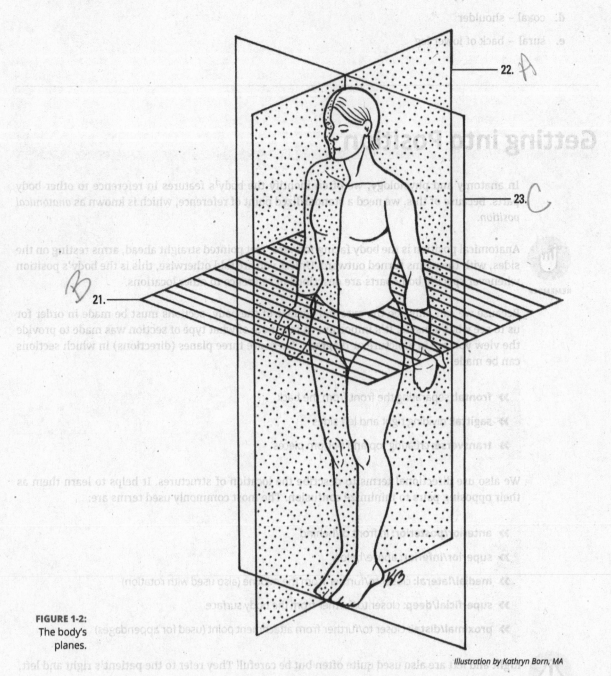

22. A

23. C

21. B

FIGURE 1-2:
The body's
planes.

Illustration by Kathryn Born, MA

24-28

Fill in the blanks.

24 The neck is _____ to the hips.

25 The lungs are _____ to the rib cage.

26 The nose is _____ to the ears.

27 The wrist is _____ to the shoulder.

28 The buttocks are _____ to the navel (belly button).

24. SUPERIOR

25. DEEP

26. medial

27. Distal

28. POSTERIOR

Answers to Questions on Terminology

The following are answers to the practice questions presented in this chapter.

(1-10) Figure 1-1 should be labeled as follows:

1. **j.** ventral, 2. **d.** dorsal, 3. **i.** thoracic, 4. **b.** abdominopelvic, 5. **c.** cranial, 6. **h.** spinal, 7. **g.** pleural, 8. **f.** pericardial, 9. **a.** abdominal, 10. **e.** pelvic

(11) The outermost layer encasing the heart: **a. parietal pericardium**

(12) The membrane that lies on the surface of the liver: **e. visceral peritoneum**

(13) The surface of the heart: **d. visceral pericardium**

(14) The lining of the thoracic cavity: **c. parietal pleura**

(15) The membrane making direct contact with the lungs: **f. visceral pleura**

(16) The layer that lines the abdominopelvic cavity: **b. parietal peritoneum**

TIP

Don't memorize all nine terms (cavities included), memorize the naming system. The space is always the cavity and the visceral layer is always making direct contact with an organ. The pattern holds true everywhere (except for surrounding the brain and spinal cord; they're special).

(17) The cephalic region is considered part of the appendicular body. **False.** The cephalic region is the head and though it does stick off the trunk, it's axial. Only the arms and legs are appendicular.

(18) Which body part would be affected if you injured your tarsal region? **c. ankle**

(19) If you suffered a laceration (cut) to your chin, the injury would be located in the **d. mental region.**

(20) Identify the correct pairing of terms: **e. sural – back of lower leg**

(21-23) Figure 1-2 should be labeled as follows: **21. b.** transverse, **22. a.** sagittal, **23. c.** frontal

(24) The neck is <u>superior</u> to the hips.

(25) The lungs are <u>deep</u> to the ribcage.

(26) The nose is <u>medial</u> to the ears.

(27) The wrist is <u>distal</u> to the shoulder.

(28) The buttocks are <u>posterior</u> to the navel (belly button).

Chapter **2**

The Chemistry of Life

We can hear your cries of alarm. You thought you were getting ready to learn about the knee bone connecting to the thigh bone. How in the heck does that involve (horrors!) *chemistry?* As much as you may not want to admit it, chemistry — particularly *organic chemistry*, the branch of the field that focuses on carbon-based molecules — is a crucial starting point for understanding how the human body works. When all is said and done, the universe boils down to two fundamental components: *matter*, which occupies space and has mass; and *energy*, the ability to do work or create change. In this chapter, we review the interactions between matter and energy to give you some insight into what you need to know to ace those early-term tests.

Building from Scratch: Atoms and Elements

All matter — be it solid, liquid, or gas — is composed of atoms. An *atom* is the smallest unit of matter capable of retaining the identity of an element during a chemical reaction. An *element* is a substance that can't be broken down into simpler substances by normal chemical reactions (the ones on the periodic table that you may have had to memorize at some point). There are 98 naturally occurring elements in nature (though 10 of these have only ever been observed in trace amounts) and 20 (at last count) artificially created elements for a total of 118 known elements. The periodic table of elements organizes all the elements by name, symbol, atomic weight, and atomic number. The *bulk elements* of interest to students of anatomy and physiology are

>> **Oxygen:** Symbol O

>> **Carbon:** Symbol C

>> **Hydrogen:** Symbol H

>> **Nitrogen:** Symbol N

>> **Phosphorus:** Symbol P

>> **Sulfur:** Symbol S

TIP

These six elements make up 95 percent of all living material. Just remember CHNOPS (read: chin-ops).

REMEMBER

Atoms are made up of the subatomic particles *protons* and *neutrons*, which are in the atom's *nucleus*, and clouds of *electrons* orbiting the nucleus. The *atomic weight*, or *mass*, of an atom is the total number of protons and neutrons in its nucleus. The *atomic number* of an atom is its number of protons; conveniently, atoms that are electrically neutral have the same number of positive charges as negative charges. Opposite charges attract, so negatively charged electrons are attracted to positively charged protons. The attraction holds electrons in orbits outside the nucleus. The more protons there are in the nucleus, the stronger the atom's positive charge is and the more electrons it can attract.

>> The first shell holds only two electrons.

>> The second and third shells hold eight electrons each.

>> The fourth shell (which can be found in elements such as potassium, calcium, and iron) holds up to 18 electrons. Higher shells also exist that hold even more electrons.

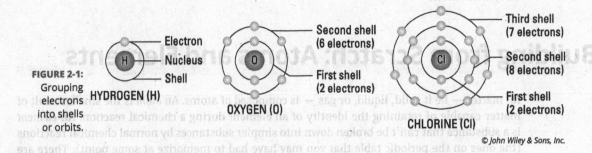

FIGURE 2-1: Grouping electrons into shells or orbits.

© John Wiley & Sons, Inc.

Other key chemistry terms that you need to know are

>> **Isotopes:** Atoms of an element that have a different number of neutrons and a different atomic weight than usual. In other words, isotopes are alternate forms of the same chemical element, so they always have the same number of protons as that element but a different number of neutrons. The two most common in the body are potassium 40 and carbon 14.

>> **Ions:** Because electrons are relatively far from the atomic nucleus, they are most susceptible to external fields. Atoms that have gained or lost electrons are transformed into ions. Getting an extra electron turns an atom into a negatively charged ion, or *anion*, whereas losing an electron creates a positively charged ion, or *cation*. The four most prevalent are sodium (Na^+), potassium (K^+), calcium (Ca^{2+}), and chloride (Cl^-).

TIP

To keep anions and cations straight, think like a compulsive dieter: Gaining is negative, and losing is positive.

>> **Acid:** A substance that becomes ionized when placed in solution, producing positively charged hydrogen ions, H^+. An acid is considered a proton donor. (Remember, atoms always have the same number of electrons as protons. Ions are produced when an atom gains or loses electrons.) Stronger acids separate into larger numbers of H^+ ions in solution.

>> **Base:** A substance that becomes ionized when placed in solution, producing negatively charged hydroxide ions, OH^-. Bases are referred to as being more *alkaline* than acids and are known as proton acceptors. Stronger bases separate into larger numbers of OH^- ions in solution.

>> **pH (potential of hydrogen):** A mathematical measure on a scale of 0 to 14 of the acidity or alkalinity of a substance. A solution is considered *neutral*, neither acid nor base, if its pH is exactly 7. (Pure water has a pH of 7.) A substance is *basic* if its pH is greater than 7 and *acidic* if its pH is less than 7. The strength of an acid or base is quantified by its absolute difference from that neutral number of 7. This number is large for a strong base and small for a weak base. Interestingly, skin is considered acidic because it has a pH around 5. Blood, on the other hand, is slightly basic with a pH around 7.4.

Answer these practice questions about atoms and elements:

 1 Which of the following is NOT a bulk element found in most living matter?

a. Carbon

b. Oxygen

c. Nitrogen

d. Potassium

e. Sulfur

 2 Among the subatomic particles in an atom, the two that have equal weight are

a. neutrons and electrons.

b. protons and neutrons.

c. positrons and protons.

d. neutrons and positrons.

3 For an atom with an atomic number of 19 and an atomic weight of 39, the total number of neutrons is

    a. 8.

    b. 19.

    c. 20.

    d. 39.

    e. 58.

4 Element X has 14 electrons. How many electrons are in its outermost shell?

    a. 2

    b. 4

    c. 6

    d. 8

    e. 14

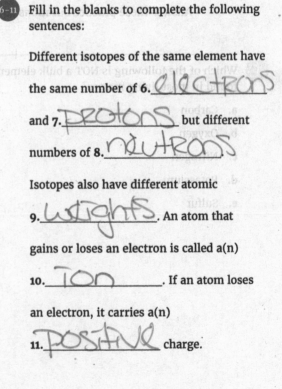

5 A substance that, in water, separates into a large number of hydroxide ions is

    a. a weak acid.

    b. a weak base.

    c. a strong acid.

    d. a strong base.

    e. neutral.

6-11 Fill in the blanks to complete the following sentences:

Different isotopes of the same element have the same number of 6. _electrons_

and 7. _protons_ but different

numbers of 8. _neutrons_

Isotopes also have different atomic

9. _weights_. An atom that

gains or loses an electron is called a(n)

10. _ion_. If an atom loses

an electron, it carries a(n)

11. _positive_ charge.

Chemical Reactions

In the following sections, we describe the chemical bonds that hold together molecules and the organic compounds created by *chemical reactions*. Chemical reactions are at the root of many of our physiological processes. A chemical reaction rearranges the atoms of the *reactant* molecules to generate new, *product* molecules. The three most common chemical reactions in our bodies are:

REMEMBER

>> **Synthesis:** A + B → AB; for example, the formation of proteins

>> **Decomposition:** AB → A + B; for example, nutrient digestion

>> **Exchange:** AB + CD → AC + BD; for example, buffers counteracting a pH change

Chemical bonds

Atoms tend to arrange themselves in the most stable patterns possible, which means that they have a tendency to complete or fill their outermost electron orbits. They join with other atoms to do just that. The force that holds atoms together in collections known as *molecules* is referred to as a *chemical bond*. Atoms of different elements have varying affinities for electrons, measured by their *electronegativity*. This dictates the type of chemical bonds they will form, which are described in the next few sections.

Ionic bond

This chemical bond (shown in Figure 2-2) involves a transfer of an electron to the atom with higher electronegativity. So one atom gains an electron while one atom loses an electron. One of the resulting ions carries a negative charge (anion), and the other ion carries a positive charge (cation). Because opposite charges attract, the atoms bond together to form a molecule.

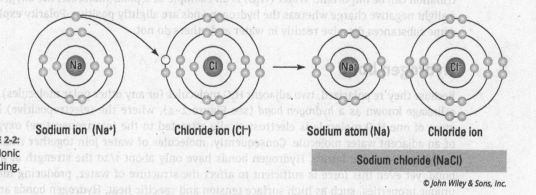

Sodium ion (Na⁺)   Chloride ion (Cl⁻)   Sodium atom (Na)   Chloride ion

FIGURE 2-2:
Ionic
bonding.

Sodium chloride (NaCl)

© John Wiley & Sons, Inc.

Covalent bond (non-polar)

The most common bond in organic molecules, a covalent bond (shown in Figure 2–3) involves the sharing of electrons between two atoms with similar electronegativities. In the human body, this is the most stable connection between atoms (though this does not hold true outside of the presence of water).

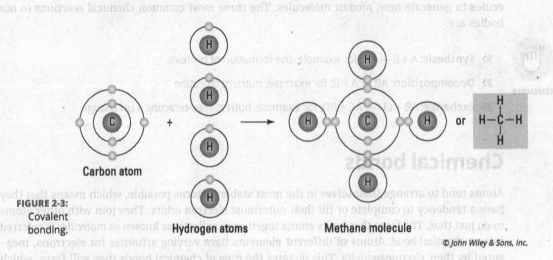

© John Wiley & Sons, Inc.

FIGURE 2-3:
Covalent
bonding.

Carbon atom

Hydrogen atoms

Methane molecule

Polar covalent bond

Sometimes atoms with differing electronegativity will form a covalent bond, producing an unevenly distributed charge. This is known as a *polar bond* (shown in Figure 2-4), an intermediate case between ionic and covalent bonding, with one end of the molecule slightly negatively charged and the other end slightly positively charged. These slight imbalances in charge distribution are indicated in Figure 2-4 by lowercase delta symbols (δ) with a charge superscript (+ or −). Although the resulting molecule is neutral, at close distances the uneven charge distribution can be important. Water (H_2O) is an example of a polar molecule; the oxygen end has a slight negative charge whereas the hydrogen ends are slightly positive. Polarity explains why some substances dissolve readily in water and others do not.

Hydrogen bond

Because they're polarized, two adjacent H_2O molecules (or any other polar molecules) can form a linkage known as a *hydrogen bond* (see Figure 2-4), where the (electropositive) hydrogen atom of one H_2O molecule is electrostatically attracted to the (electronegative) oxygen atom of an adjacent water molecule. Consequently, molecules of water join together transiently in a hydrogen-bonded lattice. Hydrogen bonds have only about 1/20 the strength of a covalent bond, yet even this force is sufficient to affect the structure of water, producing many of its unique properties, such as high surface tension and specific heat. Hydrogen bonds are important in many life processes, such as in replication and defining the shape of DNA molecules.

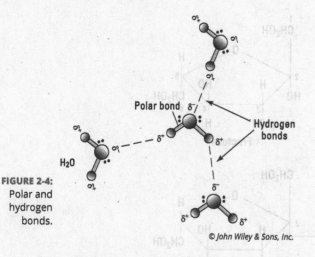

Polar bond

Hydrogen
bonds

H_2O

FIGURE 2-4:
Polar and
hydrogen
bonds.

© John Wiley & Sons, Inc.

Organic compounds

REMEMBER

When different elements combine through chemical reactions, they form *compounds.* When compounds contain carbon, they're called *organic compounds.* The four families of organic compounds with important biological functions are covered in the following sections.

Carbohydrates

These molecules consist of carbon, hydrogen, and oxygen in a ratio of roughly 1:2:1.

TIP

If a test question involves identifying a compound as a carbohydrate, count the atoms and see if they fit that ratio.

Carbohydrates are formed by the chemical reaction process of *condensation,* or *dehydration synthesis,* and broken apart by *hydrolysis,* the cleavage of a compound by a reaction that adds water. There are several subcategories of carbohydrates:

>> *Monosaccharides,* or *simple sugars,* are the building blocks, or *monomers,* of larger carbohydrate molecules and are a source of stored energy (see Figure 2-5). Key monomers include *glucose, fructose,* and *galactose.* These three have the same numbers of carbon (6), hydrogen (12), and oxygen (6) atoms in each molecule — formally written as $C_6H_{12}O_6$ — but the bonding arrangements are different. Molecules with this kind of relationship are called *isomers.* Two important five-carbon monosaccharides (pentoses) are *ribose,* a component of ribonucleic acids (RNA), and *deoxyribose,* a component of deoxyribonucleic acids (DNA).

>> *Disaccharides* are sugars formed by the bonding of two monosaccharides, including *sucrose* (table sugar), *lactose,* and *maltose.*

>> *Oligosaccharides* (from the Greek *oligo,* a few, and *sacchar,* sugar) contain three to nine simple sugars that serve many functions. They are found on plasma membranes of cells where they function in cell-to-cell recognition.

>> *Polysaccharides* are *polymers,* formed when many monomers bond into long, chainlike molecules. *Glycogen* is the primary polymer in the body; it breaks down into individual monomers of glucose, which cells use to generate usable energy.

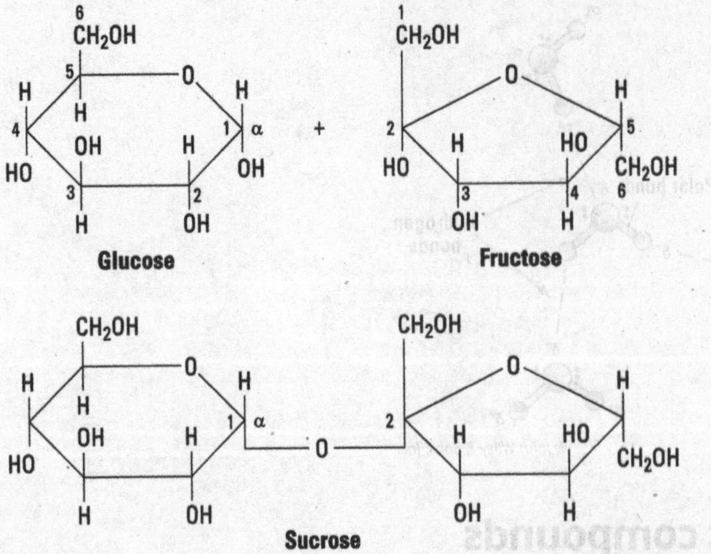

FIGURE 2-5: Monosaccharides.

Glucose

Fructose

Sucrose

© John Wiley & Sons, Inc.

Lipids

The most commonly known lipids are fats. These molecules consist of a 3-carbon *glycerol* linked to *fatty acid chains*. Insoluble in water because they contain an abundance of nonpolar bonds, lipid molecules have six times more stored energy than carbohydrate molecules. Upon hydrolysis, however, most fats form glycerol and fatty acids. A fatty acid is a long, straight chain of carbon atoms with hydrogen atoms attached (see Figure 2-6). If the carbon chain has its full number of hydrogen atoms, the fatty acid is *saturated* (examples include butter and lard). If the carbon chain has less than its full number of hydrogen atoms due to double bonds, the fatty acid is *unsaturated* (examples include margarine and vegetable oils). *Phospholipids*, as the name suggests, contain phosphorus and often nitrogen in place of one fatty acid chain. These are aligned side-by-side to form the cell membrane. Other lipids include cholesterol, vitamins A and D, and the *steroid hormones*.

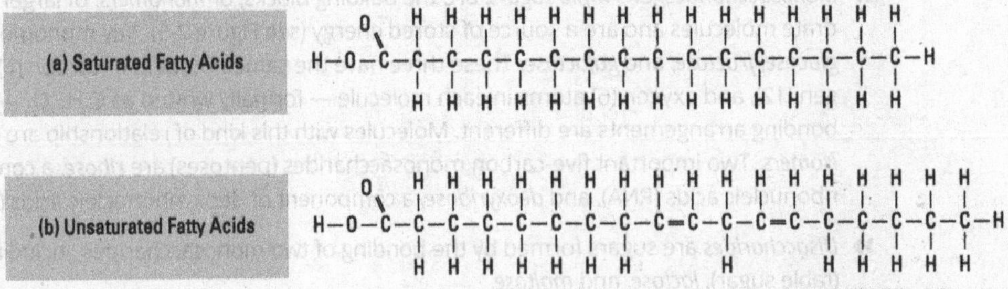

(a) Saturated Fatty Acids

(b) Unsaturated Fatty Acids

FIGURE 2-6: Fatty acids.

© John Wiley & Sons, Inc.

Proteins

Among the largest molecules, proteins can reach molecular weights of some 40 million atomic units. Proteins always contain hydrogen, oxygen, nitrogen, and carbon and sometimes contain phosphorus and sulfur. Examples of proteins in the body include *antibodies*, hemoglobin (the red pigment in red blood cells), and *enzymes* (catalysts that accelerate reactions in the body).

The human body builds protein molecules using 20 different kinds of monomers called *amino acids* (see Figure 2-7). An amino acid is a carbon atom attached to a hydrogen atom, an *amino group* (-NH₂), a *carboxyl group* (-COOH), and a unique side chain called the *R group*. Amino acids link together by *peptide bonds* to form long molecules called *polypeptides*, which then assemble into proteins. These bonds form when the carboxyl group of one molecule reacts with the amino group of another molecule, releasing a molecule of water (*dehydration synthesis*). A polypeptide, however, is not a functioning protein. It must then be folded, twisted, and often linked with other polypeptides to create a three-dimensional structure which allows it to carry out its function.

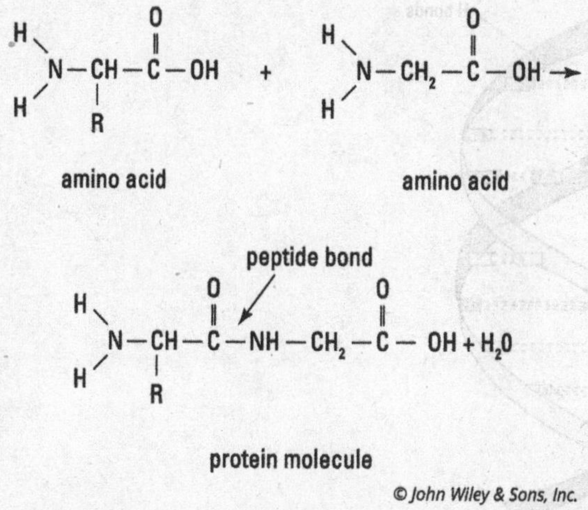

FIGURE 2-7:
Amino acids in a protein molecule.

© *John Wiley & Sons, Inc.*

Nucleic acids

These long molecules, found primarily in the cell's nucleus, act as the body's genetic blueprint. They're comprised of smaller building blocks called *nucleotides*. Each nucleotide, in turn, is composed of a five-carbon sugar (*deoxyribose* or *ribose*), a phosphate group, and a nitrogenous base. The sugar and phosphate groups link to form the backbone of the molecule. The base is attached to the sugar and aligns with its partner on the other strand (see Figure 2-8). The nitrogenous bases in DNA (deoxyribonucleic acid) are *adenine, thymine, cytosine,* and *guanine;* they always pair off A–T and C–G forming hydrogen bonds between the bases, creating the rungs of the DNA ladder. In RNA (ribonucleic acid), which occurs in a single strand, thymine is replaced by *uracil,* so the nucleotides pair off A–U and C–G during transcription (see Chapter 3 for more on this).

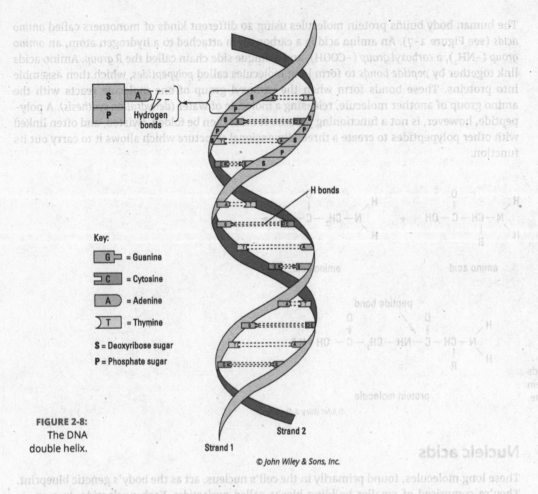

Key:

G = Guanine

C = Cytosine

A = Adenine

T = Thymine

S = Deoxyribose sugar

P = Phosphate sugar

H bonds

Hydrogen bonds

FIGURE 2-8:
The DNA
double helix.

Strand 1

Strand 2

© John Wiley & Sons, Inc.

The following is an example question dealing with chemical reactions:

EXAMPLE

Q. Oxygen can react with other atoms because it has

a. two electrons in its inner orbit.

b. eight protons.

c. an incomplete outer electron orbit.

d. eight neutrons.

A. The correct answer is an incomplete outer electron orbit. Even if you don't know the first thing about oxygen, remembering that atoms tend toward stability answers this question for you.

12 Covalent bonds are a result of

a. completing an inner electron orbit.

b. sharing one or more electrons between atoms.

c. deleting neutrons from both atoms.

d. reactions between protons.

e. transferring electrons to another atom.

13 The formation of chemical bonds is based on the tendency of an atom to

a. move protons into vacant electron orbit spaces.

b. fill its outermost energy level.

c. radiate excess neutrons.

d. pick up free protons.

14–19 Fill in the blanks to complete the following:

Molecules like water are 14. *polar*

because they share their electrons unequally.

This is due to oxygen having a higher

15. *negative* than hydrogen. As a

result, the oxygen side of the molecule has a

16. *negative* charge while the

hydrogen sides have a 17. *positive*

charge. When exposed to other polar mole-

cules 18. *opposite* charges attract and

they form 19. *hydrogen* bonds.

20 Which of the following statements is *not* true of DNA?

a. DNA is found in the nucleus of the cell.

b. DNA can be replicated.

c. DNA contains the nitrogenous bases adenine, thymine, guanine, cytosine, and uracil.

d. DNA forms a double-helix molecule.

e. DNA is made of monomers called nucleotides.

21 Polysaccharides

    a. can be reduced to fatty acids.

    b. contain nitrogen and phosphorus.

    c. are complex carbohydrates.

    d. are monomers of glucose.

    e. contain adenine and uracil.

22 Amino acids

    a. help reduce carbohydrates.

    b. are the building blocks of proteins.

    c. modulate the production of lipids.

    d. catalyze chemical reactions.

    e. control nucleic acids.

Cycling through Life: Metabolism

Metabolism (from the Greek *metabole*, which means "change") is the term for all the chemical reactions our body must undergo to sustain life. Proteins called *enzymes* control the rate of these reactions that are of two types:

>> **Anabolic reactions** require a source of energy to build up compounds that the body needs.

>> **Catabolic reactions** break down larger molecules into smaller ones, releasing energy (Memory tip: It can be *cata*strophic when things break down).

The chemical alteration of molecules in the cell is referred to as *cellular metabolism*. Enzymes are biological catalysts that accelerate chemical reactions without being changed. Thus, one way we can control the rate of our cellular metabolism is through the production of enzymes (or lack of production). In this section, we discuss the processes that take energy from the food we eat and turn it into energy that a cell can actually use. To be fair, there are numerous other metabolic reactions occurring as well — like the breakdown of wastes in the liver and the building of structural proteins — but those are discussed in their relevant chapters.

Before we dive into the energy from food, we first need to discuss the energy that cells can use — a nucleic acid called *adenosine triphosphate* (ATP). As the *tri*– prefix implies, a single molecule of ATP is composed of three phosphate groups attached to adenosine (a nitrogenous base of adenine paired with ribose). ATP's energy is stored in the bonds that attach the second and third phosphate groups. (The high-energy bond is symbolized by a wavy line in Figure 2-9.) When a cell needs energy, it removes one or two of these phosphate groups, releasing energy and converting ATP into either the two-phosphate molecule *adenosine diphosphate* (ADP) or the one-phosphate molecule *adenosine monophosphate* (AMP). Later, through additional metabolic reactions, the second and third phosphate groups are reattached to adenosine, reforming an ATP molecule until energy is needed again. Thus, ATP works like a rechargeable battery. (*Note*: Most often, we cycle through ADP and ATP; AMP is only utilized in special scenarios.)

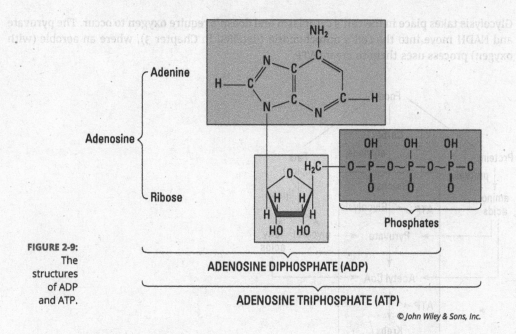

FIGURE 2-9: The structures of ADP and ATP.

© John Wiley & Sons, Inc.

Energy from food

ATP is created via *cellular respiration*, a process that occurs in and around the mitochondria of a cell (you can see this entire process outlined in Figure 2-10). This is a series of reactions that includes glycolysis, the Krebs cycle, and the electron transport chain. It starts with a single molecule of glucose and produces 38 molecules of ATP. This process, though, requires energy sources of its own, which we gain from the foods we eat. The most efficient source (glucose) is gained from *carbohydrate metabolism*. When that is used up, we turn to the products of *lipid metabolism* and finally to the products of *protein metabolism*. Each energy source enters the cellular respiration process at different points (shown in Figure 2-10).

Carbohydrate metabolism

The metabolism of all carbohydrates leads to the eventual production of glucose. The glucose is then utilized in the first phase of cellular respiration, *glycolysis*.

From the Greek *glyco* (sugar) and *lysis* (breakdown), glycolysis is the process that takes a glucose (six carbons) molecule and splits it into two molecules of *pyruvate* (three carbons each). The pyruvate, or *pyruvic acid*, will then enter the Krebs cycle to finish carrying out cellular respiration. This process, called phosphorylation, requires the energy from two molecules of ATP. However, by the end of the process we have a net gain of two ATP. In addition to the two pyruvates produced, it also yields two molecules of NADH (another energy molecule), both of which will be used later in cellular respiration.

Glycolysis takes place in the cell's cytoplasm and doesn't require oxygen to occur. The pyruvate and NADH move into the cell's mitochondria (detailed in Chapter 3), where an aerobic (with oxygen) process uses them to create ATP.

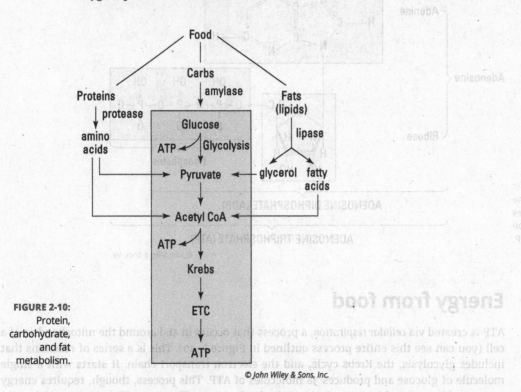

FIGURE 2-10:
Protein, carbohydrate, and fat metabolism.

© John Wiley & Sons, Inc.

Lipid metabolism

Lipids contain about 99 percent of the body's stored energy and can be digested at mealtime, but as people who complain about fats going "straight to their hips" can attest, lipids are more inclined to be stored in *adipose tissue* — the stuff generally identified as body fat. When the body is ready to metabolize lipids, an enzyme called *lipase* splits them into their monomer units: glycerol and fatty acid chains. Glycerol, which has three carbons, is then converted into pyruvate. A different series of catabolic reactions breaks apart two carbon atoms from the end of a fatty acid chain to form acetyl CoA, which then enters the Krebs cycle to produce ATP. Those reactions continue to strip two carbon atoms at a time until the entire fatty acid chain is converted into acetyl CoA molecules.

Protein metabolism

Protein metabolism focuses on producing the amino acids needed for synthesis of protein molecules within the body. But, like the lipids, the amino acids can be converted into pyruvate and acetyl CoA. Additionally, protein metabolism generates energy molecules such as NADH and $FADH_2$ that power the final step of cellular respiration — the electron transport chain.

Getting energy from energy

Now we take a look at how we can utilize the metabolites of carbohydrate, lipid, and protein metabolism to generate ATP, the cell's energy currency. This occurs through the final two steps of cellular respiration: the Krebs cycle and the electron transport chain (the first step, glycolysis, was discussed in the carbohydrate metabolism section). You can see the entire process detailed in Figure 2-11.

Krebs cycle

Also known as the *citric acid cycle*, this series of energy-producing chemical reactions begins in the mitochondria after pyruvate arrives. Before the Krebs cycle can begin, the pyruvate loses a carbon dioxide group to form *acetyl coenzyme A* (acetyl CoA). Then acetyl CoA combines with a four-carbon molecule (*oxaloacetic acid*, or OAA) to form a six-carbon citric acid molecule that then enters the Krebs cycle. The CoA is released intact to bind with another acetyl group, reforming the acetyl CoA to be used with the next pyruvate.

The cycle goes through eight steps, rearranging the atoms of citric acid to produce different intermediate molecules called *keto acids*. This generates one ATP (for each pyruvate) and four energy molecules (three NADH and one $FADH_2$) which power the electron transport chain. In the final step of the Krebs cycle, OAA is regenerated to get the next cycle going, and the carbon dioxide produced during this cycle is released from the cell to be exhaled from the lungs.

Electron transport chain

The electron transport chain, which requires oxygen, is the final step of cellular respiration and generates the bulk of the ATP. The chain itself is a series of enzyme complexes aligned on the inner folds of the mitochondria. Electrons from the energy molecules NADH and $FADH_2$ are passed into the chain and bounced from one protein to the next much like a game of hot potato. At the end of the chain, a molecule of H_2O is created and the energy from the bouncing electrons power *oxidative phosphorylation* — where an enzyme called *ATP synthase* reattaches a phosphate group to ADP, creating ATP. For every glucose molecule metabolized, 34 molecules of ATP are generated by the electron transport chain.

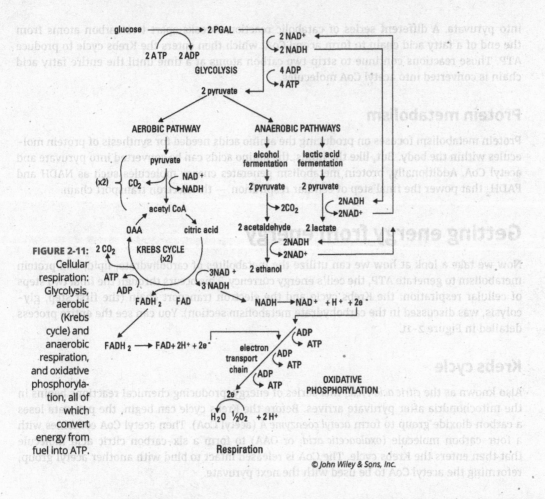

FIGURE 2-11: Cellular respiration: Glycolysis, aerobic (Krebs cycle) and anaerobic respiration, and oxidative phosphorylation, all of which convert energy from fuel into ATP.

© John Wiley & Sons, Inc.

One last thing: That severe soreness and fatigue you feel in your muscles after strenuous exercise is the result of lactic acid buildup during *anaerobic respiration* (shown in Figure 2-11). Glycolysis continues because it doesn't need oxygen to take place. But glycolysis does need a steady supply of NAD^+, which usually comes from the oxygen-dependent electron transport chain. In its absence, the body begins a process called *lactic acid fermentation*, in which one molecule of pyruvate combines with one molecule of NADH to produce ATP, a molecule of NAD^+, plus a molecule of the toxic byproduct lactic acid.

While useful for energy creation in muscles when O_2 has been used up, it hardly seems worth it — anaerobic respiration provides only two ATP molecules per glucose compared to the 38 of aerobic respiration. Tack on the creation of lactic acid, which causes *muscle fatigue* (inability of fibers to contract), as a byproduct and you may wonder why this process even exists! Well, we need to exercise, and some of the muscle fibers (but not all) will inevitably run out of oxygen. Lucky for us, the liver has a built-in mechanism to rid the body of lactic acid, though it may not happen as quickly as we'd like.

Hopefully you're not too fatigued to try out a few questions about metabolism:

Q. Cells obtain ATP by converting the energy in

  a. carbohydrates.

  b. proteins.

  c. lipids.

  d. all of these.

A. The correct answer is all of these. While it's true that carbohydrates provide the most immediately available energy, proteins and lipids also contribute to the production of ATP.

---

23 During glycolysis, a molecule of glucose is

  a. broken down to two pyruvates.

  b. broken down to two lipids.

  c. converted into energy.

  d. used to oxidize proteins.

  e. used to release oxygen.

24 By the end of cellular respiration, a single molecule of glucose can be converted into how many ATP molecules?

  a. 2

  b. 4

  c. 24

  d. 38

  e. 42

25　The buildup of lactic acid can lead to all of the following EXCEPT:

   a.　achy feeling muscles.

   b.　tired feeling muscles.

   c.　muscle cramps.

   d.　inability of fibers to contract.

26　Which two respiration processes take place in the cell's mitochondria?

   a.　Glycolysis and the Krebs cycle

   b.　Glycolysis and the electron transport chain

   c.　The Krebs cycle and the electron transport chain

   d.　The Krebs cycle and lactic acid fermentation

   e.　Lactic acid fermentation and glycolysis

27　To provide needed reactants for cellular respiration, cells utilize metabolites of the organic molecules we consume in the following order:

   a.　lipids, proteins, carbohydrates

   b.　lipids, carbohydrates, proteins

   c.　proteins, lipids, carbohydrates

   d.　carbohydrates, proteins, lipids

   e.　carbohydrates, lipids, proteins

28　What are the primary products of protein metabolism?

   a.　ATP molecules

   b.　Amino acids

   c.　Lipids

   d.　Simple sugars

   e.　Carbon dioxide molecules

Answers to Questions on Life's Chemistry

The following are answers to the practice questions presented in this chapter.

1 Which of the following is NOT a bulk element found in most living matter: **d. potassium**

2 Among the subatomic particles in an atom, the two that have equal weight are **b. protons and neutrons.** That's why you add them together to determine atomic weight, or mass.

3 For an atom with an atomic number of 19 and an atomic weight of 39, the total number of neutrons is **c. 20.** The atomic number of 19 is the same as the number of protons. The atomic weight of 39 tells you the number of protons plus the number of neutrons: 39 − 19 = 20.

4 Element X has 14 electrons. How many electrons are in its outermost shell? **b. 4.** The first orbit has the maximum two electrons, and the second orbit has the maximum eight electrons. That makes ten electrons in the first two orbits, leaving only four for the third, outermost orbit.

5 A substance that, in water, separates into a large number of hydroxide ions is **d. a strong base.** The more hydroxide ions there are, the stronger the base is.

6-11 Different isotopes of the same element have the same number of **6. electrons/protons** and **7. protons/electrons** but different numbers of **8. neutrons.** Isotopes also have different atomic **9. weights.** An atom that gains or loses an electron is called an **10. ion.** If an atom loses an electron, it carries a **11. positive** charge.

12 Covalent bonds are a result of **b. sharing one or more electrons between atoms.** If the atoms had gained or lost electrons, it would be an ionic bond, but here they're sharing — valiantly cohabiting, if you will.

13 The formation of chemical bonds is based on the tendency of an atom to **b. fill its outermost energy level.** This is true whether an atom fills its outer shell by sharing, gaining, or losing electrons.

14-19 Molecules like water are **14. polar** because they share their electrons unequally. This is due to oxygen having a higher **15. electronegativity** than hydrogen. As a result, the oxygen side of the molecule has a **16. partial negative** charge while the hydrogen sides have a **17. partial positive** charge. When exposed to other polar molecules **18. opposite** charges attract and they form **19. hydrogen** bonds.

20 Which of the following statements is *not* true of DNA? **c. DNA contains the nitrogenous bases adenine, thymine, guanine, cytosine, and uracil.** This statement is false because only RNA contains uracil.

21 Polysaccharides **c. are complex carbohydrates.** The root *poly–* means "many," which you can interpret as "complex." The root *mono–* means "one," which you can interpret as "simple."

22 Amino acids **b. are the building blocks of proteins.** Being such large molecules, proteins need to be built from complex molecules to begin with.

23 During glycolysis, a molecule of glucose is **a. broken down to two pyruvates.** Remember that glucose must become pyruvic acid before it enters the Krebs cycle.

24 By the end of cellular respiration, a single molecule of glucose can be converted into how many ATP molecules? **d. 38.** Two net molecules of ATP come from glycolysis, two molecules come from the Krebs cycle, and the electron transport chain churns out 34.

(25) The buildup of lactic acid can lead to all of the following EXCEPT: **c. muscle cramps.** Cramps are caused by sustained fiber contraction. Lactic acid buildup does just the opposite: It prevents the fiber from contracting.

(26) Which two respiration processes take place in the cell's mitochondria? **c. The Krebs cycle and the electron transport chain.** The other answers are incorrect because glycolysis takes place in the cytoplasm, and lactic acid fermentation isn't one of the three cellular respiration processes.

(27) To provide needed reactants for cellular respiration, cells utilize metabolites of the organic molecules we consume in the following order: **e. carbohydrates, lipids, proteins.** This is why it can be so difficult to burn fat by exercising, as your body will first utilize any carbohydrates that are around before breaking down that stubborn fat.

(28) The primary products of protein metabolism are **b. amino acids.** Although some ATP comes from metabolizing proteins, the body primarily needs to get amino acids from any protein that's consumed.

Chapter 3

The Cell: Life's Basic Building Block

Cytology, from the Greek word *cyto*, which means "cell," is the study of cells. Every living thing has cells, but not all living things have the same kinds of cells. *Eukaryotes* like humans (and all other organisms besides bacteria and viruses) have *eukaryotic cells*, each of which has a defined *nucleus* that controls and directs the cell's activities, and *cytosol*, fluid material found in the gel-like *cytoplasm* that fills most of the cell. Plant cells have fibrous cell walls; animal cells have a semipermeable cell membrane called a *plasma membrane* or the *plasmalemma*. Because human cells don't have cell walls, they look like gel-filled sacs with nuclei and tiny parts called *organelles* nestled inside when viewed through an electron microscope.

In this chapter, we help you sort out what makes up a cell, what all those tiny parts do, and how cells act as protein-manufacturing plants to support life's activities.

Life and Death of a Cell

At any given time, an adult has between 50–100 trillion different cells. They come in about 260 different varieties, though they all originated from that single cell in which you began your life. Pretty amazing if you stop and think about it — especially because they all have the same DNA! Cells take on their specialized structure and function through the process of *differentiation*.

Most often, chemical triggers lead to differing *gene expression;* that is, some genes are turned on while others are turned off. This is how cardiac muscle cells and nervous cells can be so dramatically different despite having the same DNA. The field of *epigenetics* strives to understand how gene expression is controlled and there is still much to be learned!

Cell differentiation is often a process with numerous intermediaries between the *stem cell* and the final, specialized one. A stem cell is an undifferentiated cell that can give rise to cells that will then specialize (or they can just make more of themselves). Only very early in embryonic development do you find truly *totipotent* stem cells — those that can develop into any cell type found in our bodies, including that of the placenta. As the embryo begins to organize, the stem cells are now *pluripotent,* which means they can develop into any cell type except the placenta. Once the embryo starts to form its layers (the endoderm, mesoderm, and ectoderm), the stem cells are now considered *multipotent.* (For more on embryonic development, refer to Chapter 17.) They have now started to follow one of three pathways: Cells in the mesoderm can develop into muscles and bones (among other things) but cannot become skin; cells in the ectoderm are on that path.

Differentiated cells then go on about their lives, carrying out their designated function until they either divide or die (though some cells like skeletal muscle and neurons share your life span so technically they do neither). Some cells, such as bone cells, for example, do not undergo mitosis. New ones must come from stem cells. Once a cell reaches its genetically determined expiration date, it begins the process of programmed cell death, or *apoptosis.* This continuous step-wise process ultimately leads to the cell shattering.

Let's see if you can sort out all of that terminology:

 Match the terms with their descriptions.

a. apoptosis

b. differentiated

c. multipotent

d. pluripotent

e. totipotent

1 ___E___ Cell that can become any type of other cell

2 ___C___ Partially differentiated stem cell; can still become numerous types but is limited

3 ___A___ Sequence of events that leads to the death of a cell

4 ___B___ Cell that carries out a specific specialized function

5 ___D___ Cell that can become most other cells, expect placenta

Gaining Admission: The Cell Membrane

Think of it as a gatekeeper, guardian, or border guard. Despite being only 6 to 10 nanometers thick and visible only through an electron microscope, the cell membrane keeps the cell's cytoplasm in place and lets only select materials enter and depart the cell as needed. This *semipermeability*, or *selective permeability*, is a result of a double layer (bilayer) of *phospholipid* molecules interspersed with protein molecules. The outer surface of each layer is made up of tightly packed *hydrophilic* (or water-loving) *polar heads*. Inside, between the two layers, you find *hydrophobic* (or water-fearing) *nonpolar tails* consisting of fatty acid chains. (See Figure 3-1.)

Since the phospholipids give the membrane a rather fluid structure, cholesterol molecules are embedded to provide stability and make it less permeable to water-soluble substances. Both cytoplasm and the *matrix*, the environment outside of the cell, are primarily water. The polar heads electrostatically attract polarized water molecules. The membrane's interior is made up of non-polar fatty acid molecules (the phospholipid tails) creating a dry middle layer. Lipid-soluble molecules can pass through this layer, but water-soluble molecules such as amino acids, sugars, and many proteins cannot. These molecules must enter the cell through little tunnels created by channel proteins. In addition to forming channels, other proteins do not span the membrane but stick out from the outer surface, sometimes with a carbohydrate attached. These serve numerous purposes: for example, creating receptors for hormones, binding sites for other cells, and aiding in recognition of self versus non-self.

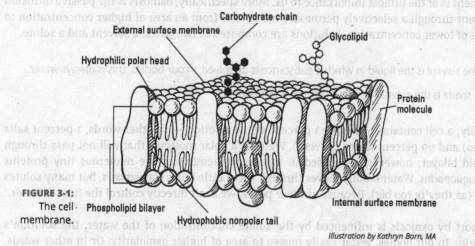

FIGURE 3-1: The cell membrane.

Illustration by Kathryn Born, MA

The cell membrane is designed to hold the cell together and to isolate it as a distinct functional unit. Although it can spontaneously repair minor tears, severe damage to the membrane will cause the cell to disintegrate. The membrane is picky about which molecules it lets in or out. It allows movement across its barrier by *diffusion*, *osmosis*, or *active transport*.

Diffusion

Atoms, ions, and molecules are in constant motion, taking a straight-line path until they inevitably crash into another particle and bounce off with a new trajectory. As a result, particles will naturally flow to areas of lower concentration (more collisions means more movement). This is *diffusion* — the movement of molecules from high to low concentration. As long as the molecules can move through any barricade that may be in their way (the cell membrane, for instance), the body doesn't need to invest any energy in their movement. The kinetic energy of the particles is the driving force moving them down their *concentration gradient*. Even once *equilibrium* has been reached (molecules are equally distributed throughout), diffusion continues; there just isn't a net change (one randomly moves out so another moves in).

So what does this have to do with cells? Diffusion is an easy way to move things into and out of a cell, passing through the cell membrane. Because it requires no energy, it is a form of *passive transport*. Small, non-polar molecules such as O_2 and CO_2 simply squeeze between phospholipid heads either entering or exiting the cell (depending on where the concentration of the molecule is lower). Since the inside of the membrane is hydrophobic, polar molecules and ions are unable to simply diffuse in (or out). They can, however, still move passively down their concentration gradient. They just require a protein channel to move through. This is termed *facilitated diffusion* as these molecules, like glucose, for example, need a protein to facilitate (allow for) their movement. Unlike the lipid-soluble molecules that use simple diffusion, we can control these water-soluble ones by opening and closing the protein channels.

Osmosis

Osmosis, simply put, is the diffusion of water. Since our bodies are 50–65 percent water, its movement is of the utmost importance to us. More specifically, osmosis is the passive diffusion of solvent through a selectively permeable membrane from an area of higher concentration to an area of lower concentration. Solutions are composed of two parts: a solvent and a solute.

>> The *solvent* is the liquid in which a substance is dissolved; in our bodies, this is always water.

>> A *solute* is the substance dissolved in the solvent.

Typically, a cell contains a roughly 1 percent saline solution — in other words, 1 percent salts (solutes) and 99 percent water (solvent). Water is a polar molecule that will not pass through the lipid bilayer; however, embedded in many cell membranes are numerous tiny proteins called *aquaporins*. Water readily moves through these little holes via osmosis, but many solutes do not (as they're too big). Because of these pores, we can't directly control the flow of water.

Transport by osmosis is influenced by the solute concentration of the water, the solution's *osmolarity*. In our bodies, water easily moves to area of higher osmolarity. Or in other words, water always goes to where there is less water. As water diffuses into a cell, *hydrostatic pressure* builds within the cell. Eventually, the pressure within the cell becomes equal to, and is balanced by, the *osmotic pressure* outside.

>> An *isotonic* solution has the same concentration of solute and solvent as found inside a cell, so a cell placed in isotonic solution — typically 1 percent saline solution for humans — experiences equal flow of water into and out of the cell, maintaining equilibrium.

>> A *hypotonic* solution has a lower solute concentration than that of the cell, which means the area with the highest osmolarity (less water) is inside the cell. If a human cell is placed in a hypotonic solution, such as distilled water, for example, water molecules move down the concentration gradient into the cell. This causes the cell to swell and could continue until the cell membrane bursts.

>> A *hypertonic* solution has a higher osmolarity outside of the cell. So the membrane of a human cell placed in 10 percent saline solution (10 percent salt and 90 percent water) would let water flow out of the cell (to where osmotic pressure is lower because water concentration is lower), therefore causing it to shrivel.

Active transport

Often, our bodies want to maintain an imbalance of concentration inside versus outside of a cell. This is particularly true for ions: Their imbalance is the foundation of nervous physiology, and our ability to move these ions allows us to influence osmosis. Since ions, for example, are being pumped against their concentration gradient, this movement requires energy. *Active transport* occurs across a semipermeable membrane, moving from the area of lower concentration to the area of higher concentration. Active transport also lets cells obtain nutrients that can't pass through the membrane by other means. Embedded in the cell membrane are transmembrane proteins able to detect and move compounds through the membrane. These *carrier* or *transport* proteins use ATP to pump molecules against their concentration gradients.

Some molecules (many proteins, for example) are far too large to be transported through the cell membrane; the hole it would need to move through would cause the cytoplasm to spill out. Bringing these molecules in, *endocytosis*, or out, *exocytosis*, requires energy, but concentration gradient is irrelevant for this type of active transport. Large molecules such as hormones are manufactured in a cell but must be released in order to carry out their function. The product (hormone) is packaged in an organelle called a *vesicle*. The outer wall of the vesicle is a phospholipid bilayer, just like the cell membrane. When the vesicle reaches the cell membrane, its phospholipids align with those of the membrane. Thus, the contents of the vesicle have been released into the extracellular fluid without ever "opening" the cell. Endocytosis occurs in much the same way, just in reverse. During *pinocytosis*, a fluid droplet is pulled into the cell as the membrane pinches in. (Again, phospholipids realign never exposing the inside of the cell to the outside environment.) *Phagocytosis* is similar to pinocytosis but occurs with solid particles. When the particle makes contact with the cell, the neighboring phospholipids project outward, engulfing the particle. In order to target specific molecules, such as cholesterol, cells utilize *receptor-mediated endocytosis*. Here, the targeted molecule binds to the receptor and the membrane extends around them both, bringing them into the cell.

6-9 Fill in the blanks to complete the following sentences:

The lipid bilayer structure of the cell membrane is made possible because phospholipid molecules contain two distinct regions: The

6._____ region is attracted

to water. The 7._____

of the phospholipids thus create an internal

and external membrane surface. The

8._____ region is repelled

by water, so the 9._____

of the phospholipids create an internal layer

that will not interact with water.

10 What is a hypotonic solution?

a. A solution that has a lower concentration of water than exists in the cell

b. A solution that has a greater concentration of water than exists in the cell

c. A solution that has a lower concentration of solute inside the cell

d. A solution that has a greater concentration of solute outside the cell

e. A solution that constantly varies in concentration when compared to what exists in the cell

11 Injecting a large quantity of distilled water into a human's veins would cause many red blood cells to

a. swell and burst.

b. shrink or shrivel.

c. carry more oxygen.

d. aggregate.

e. remain as normal.

12 Which of the following processes requires energy from ATP?

    I. diffusion/osmosis

    II. endo/exocytosis

    III. facilitated diffusion

a. I and II

b. II and III

c. II only

d. III only

e. all require energy

13-16 Match the transport term to the description.

a. active transport

b. exocytosis

c. pinocytosis

d. phagocytosis

13 __C__ Droplets of liquid are brought into the cell.

14 __D__ Cell membrane extends around a molecule engulfing it into the cell.

15 __A__ Molecules are pumped from areas of low concentration to high concentration.

16 __B__ Molecules are released from the cell by a vesicle fusing with the cell membrane.

17 Transcytosis occurs in layers of cells that form barriers. One end of a cell has receptors to bind a specific molecule. The cell brings that molecule in but then immediately sends it out on the other side. Explain the two processes that make up trancytosis.

_____

_____

_____

_____

_____

Aiming for the Nucleus

The cell nucleus is the largest cellular organelle and the first to be discovered by scientists. On average, it accounts for about 10 percent of the total volume of the cell, and it holds a complete set of genes.

The outermost part of this organelle is the *nuclear envelope*, which is composed of a double-membrane barrier, each membrane of which is made up of a phospholipid bilayer. Between the two membranes is a fluid-filled space called the *perinuclear space*. The two layers fuse to form a selectively permeable barrier, but large *pores* allow relatively free movement of molecules and ions, including large proteins. Intermediate filaments lining the surface of the nuclear envelope make up the *nuclear lamina*, which functions in the disassembly and reassembly of the nuclear membrane during mitosis and binds the membrane to the endoplasmic reticulum. The nucleus also contains *nucleoplasm*, a clear viscous material that forms the matrix in which the sub-nuclear bodies are embedded.

DNA is packaged inside the nucleus in linear structures called *chromatin*. During cell division (see Chapter 4), the chromatin condenses and becomes coiled, forming *chromosomes*. Each chromosome is a DNA molecule encoded with the genetic information needed to direct the cell's activities. The most prominent subnuclear body is the *nucleolus*, a small spherical body that stores RNA molecules and produces *ribosomes*, which are exported to the cytoplasm where they translate *messenger RNA* (mRNA) to begin protein synthesis.

The following is an example question about the nucleus:

Q. Identify the only cellular organelle found within the nucleus.

    EXAMPLE

    a. lamina

    b. envelope

    c. nucleolus

    d. chromosome

A. The correct answer is nucleolus. The other options aren't organelles. True, chromosomes are found inside the cell's nucleus, but they don't serve the cell as an organ might serve the body. Remember: The suffix "*–elle*" is a diminutive that makes the word *organelle* translate into "little organ."

---

18 What is the role of the nuclear lamina?

    a. to create a fluid-filled space within the nuclear envelope

    b. to assemble and store ribosomes

    c. a matrix to support the nuclear contents

    d. a protein packaging system within the cell

    e. to break down and rebuild the envelope

19 In the nucleus of a cell that isn't dividing, DNA

    a. moves about freely.

    b. is packaged within chromatin threads.

    c. regulates the activity of mitochondria.

    d. is organized into chromosomes.

---

20 The nucleolus

    a. packages DNA.

    b. enables large molecule transport.

    c. forms a membrane around the nucleus.

    d. assembles ribosomes.

    e. supports the contents of the nucleus.

Looking Inside: Organelles and Their Functions

Molecules that pass muster with the cell membrane enter the *cytoplasm*, a mixture of macromolecules (such as proteins and RNA), small organic molecules (such as glucose), ions, and water. Because of the various materials in the cytoplasm, it's a *colloid*, or mixture of phases, that alternates from a *sol* (a liquid colloid with solid suspended in it) to a *gel* (a colloid in which the dispersed phase combines with the medium to form a semisolid material). The fluid part of the cytoplasm, called the *cytosol*, has a differing consistency based on changes in temperature, molecular concentrations, pH, pressure, and agitation.

Within the cytoplasm lies a network of fibrous proteins collectively referred to as the *cytoskeleton*. It's not rigid or permanent, but changing and shifting according to the activity of the cell. The cytoskeleton maintains the cell's shape, enables it to move, anchors its organelles, and directs the flow of the cytoplasm. The fibrous proteins that make up the cytoskeleton include the following:

>> *Microfilaments,* rodlike structures about 5 to 8 nanometers wide that consist of a stacked protein called *actin,* the most abundant protein in eukaryotic cells. They provide structural support and have a role in cell and organelle movement as well as in cell division.

>> *Intermediate filaments,* the strongest and most stable part of the cytoskeleton. They average about 10 nanometers wide and consist of interlocking proteins, including *keratin,* that are involved in maintaining cell integrity and resisting forces that pull on the cell.

>> Hollow *microtubules* about 25 nanometers in diameter and 25 to 50 microns long made of the protein *tubulin* that grow with one end embedded in the *centrosome* near the cell's nucleus. Like microfilaments, these components of cilia, flagella, and centrioles provide structural support, can shorten and elongate, and have a role in cell and organelle movement as well as in cell division.

Organelles, literally translated as "little organs," are nestled inside the cytoplasm (except for the two organelles that move, *cilia* and *flagellum,* which are found on the cell's exterior). Each organelle has different responsibilities for producing materials used elsewhere in the cell or body. Here are the key organelles and what they do:

>> **Centrosome:** Microtubules sprout from this structure, which is located next to the nucleus and is composed of two *centrioles* — arrays of microtubules — that function in separating genetic material during cell division.

>> **Cilia:** These are short, hairlike cytoplasmic projections on the external surface of the cell. In multicellular animals, including humans, cilia move materials over the surface of the cell. In some single-celled organisms, they're used for locomotion.

>> **Endoplasmic reticulum (ER):** The ER is the site of lipid and protein synthesis. This organelle makes direct contact with the cell nucleus and is composed of membrane-bound canals and cavities that aid in the transport of materials throughout the cell. The two types of ER are *rough,* which is dotted with ribosomes on the outer surface, and *smooth,* which have no

ribosomes on the surface. The rough ER manufactures membranes and secretory proteins in collaboration with the ribosomes. The smooth ER has a wide range of functions, including carbohydrate and lipid synthesis as well as breaking down drugs.

>> **Flagellum:** This whiplike cytoplasmic projection lies on the cell's exterior surface. Found in humans primarily on sperm cells where it's used for locomotion.

>> **Golgi apparatus (or body):** This organelle consists of a stack of flattened sacs with membranes. *Transport vesicles* carry newly made proteins from the ER to the golgi, and *secretory vesicles* carry the golgi's products to the cell membrane. Located near the nucleus, it functions in the storage, modification, and packaging of proteins for secretion to various destinations within the cell or for release.

>> **Lysosome:** A tiny, membranous sac containing acids and digestive enzymes, the lysosome breaks down large molecules such as proteins, carbohydrates, and nucleic acids into materials that the cell can use. The enzymes can be used to open vesicles (break down their lipid bilayer) to access material brought in via endocytosis. It also destroys foreign particles, including bacteria and viruses, and helps to remove nonfunctioning structures from the cell.

>> **Mitochondrion:** Called the "powerhouse of the cell," this usually rod-shaped organelle consists of two membranes — a smooth outer membrane and an invaginated (folded inward) inner membrane that divides the organelle into compartments. The inward folds of the inner membrane are called *cristae*. (See Figure 3-2.) The mitochondrion provides critical functions in cell respiration, which we cover in Chapter 2.

>> **Ribosomes:** These roughly 25-nanometer structures may be found along the endoplasmic reticulum or floating free in the cytoplasm. Composed of 60 percent RNA and 40 percent protein, they translate the genetic information on messenger RNA molecules to *synthesize*, or produce, a protein molecule.

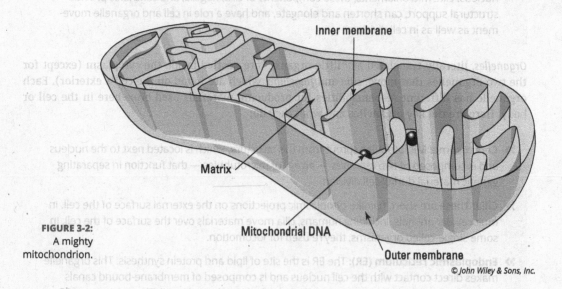

FIGURE 3-2:
A mighty mitochondrion.

© John Wiley & Sons, Inc.

Answer these practice questions about cell organelles:

21 What is the function of a mitochondrion?

a. To modify proteins, lipids, and other macromolecules

b. To produce energy through aerobic respiration

c. To keep the cell clean and recycle materials within it

d. To produce and export hormones

e. To assemble amino acids into proteins

22 Actin, which is the most abundant protein in human cells,

a. forms filaments that contribute to the cytoskeleton.

b. metabolizes lipids for cell energy.

c. prevents binding of microtubules.

d. synthesizes hormones.

e. provides the cytoplasm its fluidity.

23 Which organelle translates genetic information to direct protein synthesis?

a. The ribosome

b. The lysosome

c. The centriole

d. The nucleus

e. The vesicle

24 Which organelle is the site of protein synthesis?

a. Smooth (agranular) endoplasmic reticulum

b. Golgi apparatus

c. Nucleus

d. Rough (granular) endoplasmic reticulum

e. Mitochondria

25 Which of the following can change the consistency of cytoplasm?

a. Changes in pH

b. Temperature

c. Pressure

d. Molecule concentration

e. All of the above

26-38 Use the terms that follow to identify the cell structures and organelles shown in Figure 3-3.

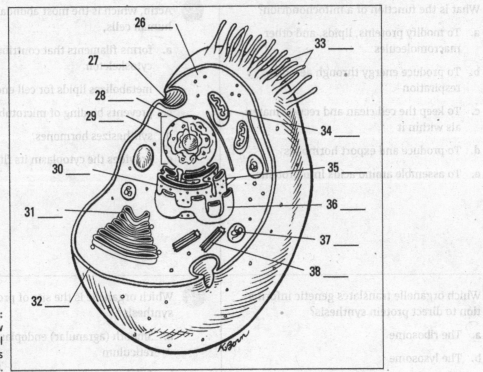

FIGURE 3-3:
A cutaway view
of an animal
cell and its
organelles.

Illustration by Kathryn Born, MA

a. Centriole

b. Cilia

c. Cytoplasm

d. Golgi apparatus

e. Lysosome

f. Mitochondrion

g. Nucleolus

h. Nucleus

i. Cell (plasma) membrane

j. Ribosomes

k. Rough endoplasmic reticulum

l. Smooth endoplasmic reticulum

m. Vesicle (formation)

Match the organelles with their descriptions.

a. Long, whiplike organelle for locomotion

b. Fluidlike interior of the cell that may become a semisolid, or colloid

c. Membranous sacs containing digestive enzymes

d. Site of cellular respiration

e. Site of ribosome production

f. Structure used for transporting molecules in, out, or around the cell

39 ___D___ Mitochondrion

40 ___E___ Nucleolus

41 ___A___ Flagellum

42 ___B___ Cytoplasm

43 ___C___ Lysosomes

44 ___F___ Vesicle

Putting Together New Proteins

Proteins are essential building blocks for all living systems, which helps explain why the word is derived from the Greek term *proteios*, meaning "holding first place." Cells use proteins to perform a variety of functions, including providing structural support and catalyzing reactions. Cells synthesize proteins through a systematic procedure that begins in the nucleus when the genetic code for a certain protein is *transcribed* from the cell's DNA into *messenger RNA*, or *mRNA*. The mRNA moves through nuclear pores to the rough endoplasmic reticulum (RER), where ribosomes *translate* the message one *codon* of three nucleotides, or *base pairs*, at a time. The ribosome uses *transfer RNA*, or *tRNA*, to fetch each required amino acid from the cytoplasm and then link them together through *peptide bonds* to form proteins (see Figure 3-4 for details).

The free ribosomes are responsible for production of proteins that remain in the cell. Proteins to be sent out are synthesized by ribosomes attached to the ER. Regardless, the linking of amino acids into an incredibly long chain does not a functional protein make. All proteins depend on their three-dimensional structure to carry out their function. Molecules called *chaperonins* aid in the folding process (either in the cytoplasm, ER, or golgi). Once a protein is assembled into its final shape (if being prepared for export), it is sorted and packaged into vesicles by the time it reaches the end of the golgi apparatus.

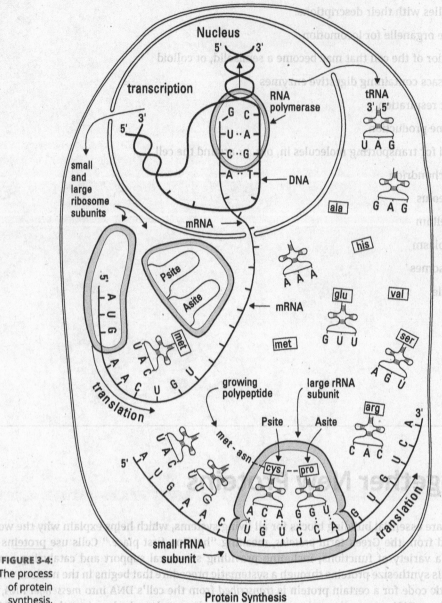

FIGURE 3-4:
The process
of protein
synthesis.

Protein Synthesis

© John Wiley & Sons, Inc.

TIP Don't let the labels confuse you. Proteins are chains of amino acids (usually very long chains of at least 100 acids). Enzymes, used to catalyze reactions, also are chains of amino acids and therefore also are categorized as proteins. *Polypeptides*, or simply *peptides*, are shorter chains of amino acids used to bond larger protein molecules, but they also can be regarded as proteins. Both antibodies and many hormones are also proteins, along with almost everything else in the body — hair, muscle, cartilage, and so on. Even the four basic blood types — A, B, AB, and O — are differentiated by the proteins (antigens) found in each.

Fill in the blanks to complete the following sentences:

Protein synthesis begins in the cell's 45. _nucleus_ when the gene for a certain

protein is 46. _transcribed_ into messenger RNA, or mRNA, which then moves on to

the 47. _Rough ER_. There, ribosomes 48. _translate_ three base

pairs at a time, forming a series also referred to as a 49. _codon_. Molecules of

50. _tRNA_ then collect each amino acid needed so that the ribosomes can link

them together through 51. _peptide_ bonds. The chain is then folded into its 3D struc-

ture with the help of 52. _chaperonins_. Now in the 53. _Golgi apparatus_ the protein

is packaged into a 54. _vesicle_ for transport to the cell membrane to be released.

---

55. Which of the following comes first in the protein-synthesis process?

a. Transfer RNA

b. Transcription

c. Peptide bonds

d. Messenger RNA

e. Translation

56. What is a codon?

a. A sequence of three adjacent nucleotides on a mRNA molecule

b. A type of amino acid used in the cellular production of protein

c. A specialized ribosome required to translate DNA

d. A base pair that connects complementary strands of mRNA and tRNA

Answers to Questions on the Cell

The following are answers to the practice questions presented in this chapter.

1. Cell that can become any type of other cell: **e. totipotent**

2. Partially differentiated stem cell; can still become numerous types but is limited: **c. multipotent**

3. Sequence of events that leads to the death of a cell: **a. apoptosis**

4. Cell that carries out a specific specialized function: **b. differentiated**

5. Cell that can become most other cells, expect placenta: **d. pluripotent**

6–9. The lipid bilayer structure of the cell membrane is made possible because phospholipid molecules contain two distinct regions: The **6. hdrophilic** region is attracted to water. The **7. heads** of the phospholipids thus create an internal and external membrane surface. The **8. hydrophobic** region is repelled by water, so the **9. tails** of the phospholipids create an internal layer that will not interact with water.

10. What is a hypotonic solution? **b. A solution that has a greater concentration of water than exists in the cell.**

TIP

The prefix *hypo* refers to under or below normal. The prefix *hyper* refers to excess, or above normal. Someone who has been out in the cold too long suffers hypothermia — literally insufficient heat. So a solution, or tonic, with very few particles would be hypotonic.

11. Injecting a large quantity of distilled water into a human's veins would cause many red blood cells to **a. swell and burst**. With a hypotonic solution outside the cell, the membrane would allow osmosis to continue past the breaking point. *Remember:* Hypo makes cells big, like a hippo.

12. Which of the following processes requires energy from ATP? **c. II only**. Endo- and exocytosis are the only ones on the list that require energy. While facilitated diffusion does require a protein for molecules to travel through, it does not need energy.

13. Droplets of liquid are brought into the cell: **c. pinocytosis**

14. Cell membrane extends around a molecule engulfing it into the cell: **d. phagocytosis**

15. Molecules are pumped from areas of low concentration to high concentration: **a. active transport**

16. Molecules are released from the cell by a vesicle fusing with the cell membrane: **b. exocytosis**

17. First, receptor mediated endocytosis occurs to bring the molecule into the cell. Since the cell membrane spreads around the molecule, it enters already packaged in a vesicle. Rather than unpack its contents, the cell simply pushes the vesicle to the opposite end where it fuses with the cell membrane, releasing the molecule via exocytosis.

18. What is the role of the nuclear lamina? **e. to break down and rebuild the envelope**

19. In the nucleus of a cell that isn't dividing, DNA **b. is packaged within chromatin threads.**

20 The nucleolus **d. assembles ribosomes.** It's not just a coincidence that the nucleolus sits at the heart of genetic storage.

21 What is the function of a mitochondrion? **b. To produce energy through aerobic respiration.** Each of the other options describes a function served by a different organelle.

22 Actin, which is the most abundant protein in human cells, **a. forms filaments that contribute to the cytoskeleton.** It makes sense that the protein making up much of the cytoskeleton is the most abundant because the cytoskeleton accounts for up to 50 percent of the cell's volume.

23 Which organelle translates genetic information to direct protein synthesis? **a. The ribosome**

When you think of a big protein-laden meal, you think of ribs. Ribs. Ribosome. Protein synthesis. Get it?

TIP

24 Which organelle is the site of protein synthesis? **d. Rough (granular) endoplasmic reticulum.** Ribosomes isn't a choice, and it's the attachment of them that makes the RER rough.

25 Which of the following can change the consistency of cytoplasm? **e. All of the above.** In addition, agitation, or just being jostled, can change cytoplasmic consistency.

26-38 Following is how Figure 3-3, the cutaway view of the cell and its organelles, should be labeled.

26. **c. Cytoplasm;** 27. **m. Vesicle (formation);** 28. **g. Nucleolus;** 29. **h. Nucleus;** 30. **k. Rough endoplasmic reticulum;** 31. **d. Golgi apparatus;** 32. **i. Cell (plasma) membrane;** 33. **b. Cilia;** 34. **f. Mitochondrion;** 35. **j. Ribosomes;** 36. **l. Smooth endoplasmic reticulum;** 37. **e. Lysosome;** 38. **a. Centriole**

39 Mitochondrion: **d. Site of cellular respiration**

40 Nucleolus: **e. Site of ribosome production**

41 Flagellum: **a. Long, whiplike organelle for locomotion**

42 Cytoplasm: **b. Fluidlike interior of the cell that may become a semisolid, or colloid**

43 Lysosomes: **c. Membranous sacs containing digestive enzymes**

44 Vesicle: **f. Structure used for transporting molecules in, out, or around the cell**

45-54 Protein synthesis begins in the cell's **45. nucleus** when the gene for a certain protein is **46. transcribed** into messenger RNA, or mRNA, which then moves on to the **47. rough endoplasmic reticulum.** There, ribosomes **48. translate** three base pairs at a time, forming a series also referred to as a **49. codon.** Molecules of **50. tRNA** then collect each amino acid needed so that the ribosomes can link them together through **51. peptide** bonds. The chain is then folded into its 3D structure with the help of **52. chaperonins.** Now in the **53. Golgi apparatus** the protein is packaged into a **54. vesicle** for transport to the cell membrane to be released.

55 Which of the following comes first in the protein synthesis process? **b. Transcription.** Remember that you have to transcribe before you can translate and the result of transcription is mRNA.

56 What is a codon? **a. A sequence of three adjacent nucleotides on a mRNA molecule.** It takes three nucleotides to round up a single amino acid.

Chapter 4

Divide and Conquer: Cellular Mitosis

E ver had so many places to be that you wished you could just divide yourself in two? If only we could be more like the cells that make us up! Cell division is how one "mother" cell becomes two identical twin "daughter" cells. Cell division takes place for several reasons:

» **Growth:** Multicellular organisms, humans included, each start out as a single cell — the fertilized egg. That one cell divides (and divides and divides), eventually becoming an entire complex being.

» **Injury repair:** Uninjured cells in the areas surrounding damaged tissue divide to replace those that have been destroyed.

» **Replacement:** Cells eventually wear out and cease to function. Their younger, more functional neighbors divide to take up the slack.

» **Asexual reproduction:** No, human cells don't do this. Only single-celled organisms do.

Cell division occurs over the course of two processes: *mitosis*, which is when the chromosomes within the cell's nucleus duplicate to form two daughter nuclei; and *cytokinesis*, which takes place when the cell's cytoplasm divides to surround the two newly formed nuclei. Although cell division breaks down into several stages, there are no pauses from one step to another. Cell division as a whole is often referred to as mitosis because most of the changes occur during that process. Cytokinesis doesn't start until later, but mitosis and cytokinesis do end together.

Keep in mind: Cells are living things, so they mature, reproduce, and die. They are not always actively engaged in the reproduction part. In this chapter, we review the entire cell cycle, and you get plenty of practice figuring out what happens when and why.

Walking through the Cell Cycle

The time that precedes cell division is referred to as *interphase*, and the majority of a cell's life is spent here. The length of time is highly variable as our cells have drastically different programmed life spans. Certain types of white blood cells and cells that line parts of our digestive tract often don't live to see 24 hours. Others, such as neurons and cardiac muscle cells, have the same life span as you. Regardless, each cell starts its life at the beginning of interphase and must go through two checkpoints before beginning the division process. After living part of their useful lives carrying out their specialized function, they reach the *restriction checkpoint*. From here, there are three options — though the choice is genetically programmed:

>> **Undergo apoptosis:** *Apoptosis*, or programmed cell death, is the fate of cells that have reached their life span (like an expiration date). Enzymes trigger this process and the cell is dismantled from the inside out (though it looks more like they're exploding).

>> **Remain specialized:** Cells continue to carry out their functions but do not divide. There's usually no going back from here; the cells live but never divide. This is the case for neurons, most muscle cells, and most white blood cells, among others.

>> **Prepare for division:** Interphase does not end here. There's still work to be done before cell division can begin.

Waiting for action: Interphase

Interphase is the period when the cell isn't dividing. It begins when the new cells are done forming and ends when the cell is prepared to divide. Although it's also called a "resting stage," there's constant activity in the cell during interphase, which is broken down into three subphases:

>> **G$_1$:** Stands for "growth" or "gap." During G$_1$, the cell grows and develops, begins metabolism, and synthesizes proteins. The restriction checkpoint marks the end of this phase.

>> **S:** Stands for "synthesis." *DNA synthesis* or *replication* occurs during this subphase. Each double-helix DNA molecule inside the cell's nucleus is copied, creating two *sister chromatids*. At the end of S phase, the cell has two copies of every piece of DNA (chromosome) so that each daughter cell can receive a full set.

>> **G$_2$:** Stands for "gap." Enzymes and proteins needed for cell division are produced during this subphase. The cell continues to grow, more rapidly this time, and often additional organelles are made to dole out to the daughter cells. The second checkpoint occurs here, halting the process if DNA damage is detected. Otherwise, the cell officially begins the division process, or *mitosis*.

Sorting out the parts: Prophase

As the first phase of mitosis, *prophase* is when structures of the cell's nucleus begin to disappear, including the nuclear membrane (or envelope), nucleoplasm, and nucleoli. The two centrosomes (duplicated during S phase), each containing two centrioles, push apart to opposite ends of the cell, forming poles. The centrioles produce protein filaments that form *spindle fibers* between the poles as well as *asters* (or *astral rays*) that radiate from the poles into the cytoplasm.

At the same time, the *chromatin* threads (or pieces of DNA) shorten and coil, forming *chromonemata*. These wrap around proteins called *histones* to further condense the structure forming visible *chromatids*. The two identical chromatids are linked in the middle by a *centromere* creating the x-shaped *chromosome* you're familiar with. Now the DNA is more manageable to move around and distribute equally between the two cells.

Dividing at the equator: Metaphase

Now that the nucleus is gone, the chromosomes migrate to the middle of the cell. They line up exactly along the *equatorial plane* with each centromere on the invisible center line. Lastly, the spindle fibers connect the centromeres to the centrioles on the opposite ends of the cell.

Packing up to move out: Anaphase

In anaphase, the centromeres are split apart by the centrioles reeling in the spindle fibers. The sister chromatids (identical copies of DNA) are pulled to the opposite poles. The cell begins to elongate and a slight furrow develops in the cytoplasm, showing where cytokinesis will eventually take place.

Pinching off: Telophase

Telophase occurs as the chromosomes reach the poles and the cell nears the end of division. The spindles and asters of early mitosis disappear, and each newly forming cell begins to synthesize its own nucleus to enclose the separated chromosomes. The coiled chromosomes unwind, becoming chromatin once again. There's a more pronounced pinching, or furrowing, of the cytoplasm into two separate bodies, but there continues to be only one cell.

Splitting up: Cytokinesis

Cytokinesis means it's time for the big breakup. The furrow, formed by a contractile ring that will divide the newly formed sister nuclei, migrates inward until it cleaves the single, altered cell into two new cells. Each new cell is smaller and contains less cytoplasm than the mother cell, but the daughter cells are genetically identical to each other and to the original mother cell, and will grow to normal size during early interphase. Telophase is still wrapping up at this point, thus both processes end at the same time.

Try remembering the phases like this:

TIP

Interphase = intermission, Prophase = prepare, Metaphase = middle, Anaphase = apart, Telophase = twins, Cytokinesis = complete

Now try this warm-up question:

Q. True or false: Cells are dormant during interphase.

A. False. While it's also called a "resting stage" (a misnomer!), plenty is going on during interphase.

---

1 The G_1 subphase of interphase is

a. the period of DNA synthesis.

b. the most active phase.

c. the phase between S and G_2.

d. the phase immediately before mitosis begins.

e. part of cell division.

2 What happens to DNA during interphase?

a. It's duplicated during the S subphase.

b. It's cut in two during the G_2 subphase.

c. It unwinds rapidly during the G_2 subphase.

d. It separates the opposite ends during the S subphase.

e. It develops a protective film during the G_1 subphase.

---

3 During prophase, the nuclear membrane, or envelope,

a. splits into thirds.

b. develops a series of concave dimples.

c. begins to disappear.

d. changes color.

e. strengthens its structure.

4 Which of the following is true for metaphase?

a. The nuclear membrane reappears.

b. The chromosomes move to the poles.

c. The chromosomes align on the equatorial plane.

d. It's composed of subphases G_1, S, and G_2.

e. Chromatin coils up making it easier to move.

5 During metaphase, each chromosome consists of two duplicate chromatids.

a. True

b. False

6 Identify an event that DOES NOT happen during anaphase.

a. Early cytokinesis occurs with slight furrowing.

b. The cell goes through the growth subphase.

c. Centrioles shorten the spindles fibers.

d. Chromatids move to the outside of the cell.

e. The centromeres split.

7 Which event DOES NOT occur during telophase?

a. The chromosomes uncoil.

b. The chromosomes reach the poles.

c. The spindle fibers degenerate.

d. The nuclear membrane reforms.

e. The cell splits in two.

8 What structures disappear during telophase?

a. Spindles and asters

b. Nuclear membranes

c. Nucleolei

d. Cytoplasm

e. Chromatin

9 Which is the correct order of phases of the cell cycle and mitosis?

a. Prophase, metaphase, anaphase, telophase, interphase, cytokinesis

b. Interphase, cytokinesis, telophase, prophase, anaphase, metaphase

c. Cytokinesis, interphase, telophase, anaphase, metaphase, prophase

d. Interphase, prophase, metaphase, anaphase, telophase, cytokinesis

e. Prophase, metaphase, cytokinesis, anaphase, telophase, metaphase

10-21   Use the terms that follow to identify the stages and cell structures shown in Figure 4-1.

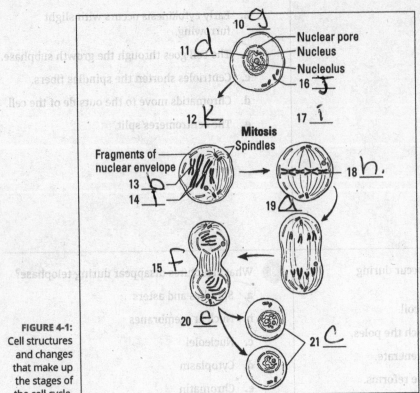

10 __g__

11 __d__   Nuclear pore
             Nucleus
             Nucleolus
16 __j__

12 __k__   17 __i__

             Mitosis
             Spindles

Fragments of
nuclear envelope
13 __b__   18 __h__
14 __l__

                   19 △

15 __f__

20 __e__   21 __c__

FIGURE 4-1:
Cell structures
and changes
that make up
the stages of
the cell cycle.

Illustration by Kathryn Born, MA

a. Anaphase

b. Centromere

c. Daughter cells

d. Chromatin

e. Cytokinesis

f. Telophase

g. Interphase

h. Chromosomes aligned at equator

i. Metaphase

j. Centriole

k. Prophase

l. Sister chromatids

22 The two newly formed daughter cells are

a. the same size as the mother cell.

b. not genetically identical to each other.

c. unequal in size.

d. genetically identical to the mother cell.

23 Cytokinesis can be described as

a. the period of preparation for cell division.

b. the dividing of the cytoplasm to surround the two newly formed nuclei.

c. the stage of alignment of the chromatids on the equatorial plane.

d. the initiation of cell division.

e. the dissolution of the cell's nucleus.

Understanding What Can Go Wrong during Cell Division

Mitosis is said to be a continuous process. That is, once it starts it doesn't stop; cells proceed through all the phases automatically. Though there are checkpoints throughout (and even proofreaders to check DNA replication during S phase), some errors are inevitably missed. With the millions upon millions of cell divisions that happen in the human body, it's not surprising that sometimes things go wrong.

An error during DNA replication is referred to as a *gene mutation*. An error during mitosis leads to a *chromosome mutation*. As mentioned earlier, the DNA is checked for mismatched bases as it is being replicated. Mutations here can lead to a range of effects — from no discernable change, to a misfolded protein, to a completely non-functional gene. Chromosome mutations have a more drastic effect:

>> **Deletion:** Part of the chromosome just breaks off, taking the instructions of the gene with it.

>> **Insertion:** Additional copies of the genes are inserted into the chromosome.

>> **Inversion:** A section of the chromosome breaks off but is flipped upside down before reattaching.

>> **Nondisjunction:** An entire chromosome is pulled to the pole rather than splitting into sister chromatids. This leaves one daughter cell with three copies of each chromosome and the other with only one.

Often, in cases of chromosomal mutations especially, cells die during their first G1 phase because they are unable to perform all their necessary functions. Cells that survive can produce too much or too little of a protein affecting the relevant physiological process. If a mutation affects any of the numerous genes that control the cell cycle, mitosis can end up on fast-forward. Accelerated mitosis can lead to the formation of a tumor, also called a *neoplasm*. The rate of division usually restricts itself to replacing worn out or injured cells, but with accelerated mitosis, the cells don't know when to stop dividing.

24 What is the name of the mutation during mitosis that can result in an extra copy of chromosome 21, which causes Down syndrome?

a. Neoplasm

 b. Nondisjunction

c. Insertion

d. Inversion

e. Acceleration

 25 Accelerated mitosis can cause the growth of a tumor, which is also known as a(n)

a. nondisjunction.

b. daughter cell.

 c. neoplasm.

d. kinetochore

e. apoptosis.

 26 An error that occurs during S phase is called a chromosome mutation.

a. True

b. False

Answers to Questions on Mitosis

The following are answers to the practice questions presented in this chapter.

1. The G1 subphase of interphase is **b. the most active phase.** With all that organelle-growing, metabolizing, and protein-synthesizing that takes place during G1, this isn't surprising.

2. What happens to DNA during interphase? **a. It's duplicated during the S subphase.** Remember, the cell is S-ynthesizing new DNA molecules during this phase.

3. During prophase, the nuclear membrane, or envelope, **c. begins to disappear.** Otherwise, the spindle fibers wouldn't be able to reach the chromosomes.

4. Which of the following is true for metaphase? **c. The chromatids align on the equatorial plane.** Each of the other answer choices occurs during earlier or later phases.

5. During metaphase, each chromosome consists of two duplicate chromatids. **a. True.** That way, each resulting daughter cell will have identical chromosomes.

6. Identify an event that DOES NOT happen during anaphase. **b. The cell goes through the growth subphase.** G_1, the growth subphase, took place back in interphase.

7. Which event DOES NOT occur during telophase? **e. The cell splits in two.** This is a common mistake, especially since telophase and cytokinesis end at the same time. Remember that telophase is all about the nucleus, cytokinesis is all about the cell.

8. What structures disappear during telophase? **a. Spindles and asters.** These structures disappear because they're no longer needed at the end of mitosis.

9. Which is the correct order of phases of the cell cycle and mitosis? **d. Interphase, prophase, metaphase, anaphase, telophase, cytokinesis.**

10-21. Following is how Figure 4-1, the stages and structures of mitosis, should be labeled.

   10. g. Interphase; 11. d. Chromatin; 12. k. Prophase; 13. b. Centromere; 14. l. Sister chromatids; 15. f. Telophase; 16. j. Centriole; 17. i. Metaphase; 18. h. Chromosomes aligned at equator; 19. a. Anaphase; 20. e. Cytokinesis; 21. c. Daughter cells

22. The two newly formed daughter cells are **d. genetically identical to the mother cell.** None of the other answer options makes sense.

23. Cytokinesis can be described as **b. the dividing of the cytoplasm to surround the two newly formed nuclei.**

24. What is the name of the mutation during mitosis that can result in an extra copy of chromosome 21, which causes Down syndrome? **b. Nondisjunction.** Split the word into pieces. "Non" means negative, or it didn't happen. "Dis" means "apart" or "asunder." And you know that a "junction" is where things are joined. Down syndrome occurs when chromosomes fail to split up properly, leaving daughter cells with the wrong number of chromosomes.

25. Accelerated mitosis can cause the growth of a tumor, which is also known as a(n) **c. neoplasm.**

26. An error that occurs during S phase is called a chromosome mutation. **b. False.** S phase is when DNA is being copied, so mutations here would affect the base pairing, a gene mutation. For a chromosomal mutation, the error has to occur during mitosis when the chromosomes are condensed.

Chapter **5**

The Study of Tissues: Histology

O h, what tangled webs we weave! As the chapter title says, *histology* is the study of tissues, but you may be surprised to find out that the Greek *histo* doesn't translate as "tissue" but instead as "web." It's a logical next step after reviewing the cell and cellular division (see Chapters 3 and 4) to take a look at what happens when groups of similar cells "web" together to form tissues. The four different types of tissue in the body are as follows:

» Epithelial tissue (from the Greek *epi–* for "over" or "outer")

» Connective tissue

» Muscle tissue

» Nervous tissue

In some tissues, like many of the connective tissues, the cells do not make physical contact with each other. The space between them can be filled with various fibers or simply left "empty," though that empty space would be filled with *interstitial fluid*, which is basically water and all the molecules that are pushed out with it from the blood. (See Chapter 11 for an explanation of how this happens.) In others, such as epithelial tissues, the function is dependent on the cells being tightly packed together. This is done by *cell junctions* which come in three general varieties:

» **Tight Junctions:** Cells are basically sewn together using proteins leaving little to no space between them. These can be found in most epithelial tissues.

>> **Desmosomes or Adhesive Junctions:** Proteins between the cells connect with filaments inside of them acting much like Velcro. Cells are still bound together tightly but have more flexibility as a layer. These can be found in the epidermis (the outer layer of skin) and the lining of the uterus.

>> **Gap Junctions:** Cells are connected by proteins that form channels, allowing the two neighboring cells to directly share cytoplasm. This junction enables the cells to pass on small molecules without investing energy. These can be found in most tissues of the body but are especially important in cardiac muscle.

In this chapter, you find a quick review of the four tissue types along with practice questions to test your knowledge of them.

Getting into Your Skin: Epithelial Tissue

Perhaps because of its unique job of lining all our body surfaces (both external and internal), epithelial tissue has many characteristics that distinguish it from other tissue types.

Epithelial tissues, which are arranged in sheets of tightly packed cells, always have a free, or *apical*, surface that is exposed to the air or to fluid. The cells have a defined top and bottom and can be found in a single layer or sandwiched between other cells of the same type. Cells on the apical side sometimes have cytoplasmic projections such as *cilia*, hair-like growths that can move material over the cell's surface, or *microvilli*, fingerlike projections that increase the cell's surface area for increased adsorption and more efficient diffusion of absorbents and secretions.

 Adsorption involves the *ad*hesion of something to a surface, while *ab*sorption involves a fluid permeating or dissolving into a liquid or solid.

Opposite the apical side is the *basal* side, or bottom layer of cells. These are attached to the tissue's *basement membrane* which is a layer of non-living material secreted by the epithelial cells. It adheres to the underlying connective tissue and acts as a filter — which is especially useful since epithelial tissues lack blood flow.

Epithelial tissue serves several key functions, including the following:

>> **Protection:** Skin protects vulnerable structures or tissues deeper in the body.

>> **Barrier:** Epithelial tissues prevent foreign materials from entering the body and retain interstitial fluids.

>> **Sensation:** Sensory nerve endings embedded in epithelial tissue connect the body with outside stimuli.

>> **Secretion:** Epithelial tissue in glands can be specialized to secrete enzymes, hormones, and fluids.

>> **Absorption:** Linings of the internal organs are specialized to facilitate absorption of materials into the body.

» **Filtration:** The epithelium of the respiratory tract protects the body by filtering out dirt and particles and cleaning the air that's inhaled. The blood is filtered in epithelial tissue in the kidneys.

» **Diffusion:** Simple squamous epithelial cells form a semipermeable membrane that allows diffusion of some materials to pass through under osmotic pressure. This way, molecules diffuse between the cells rather than through them.

Single-layer epithelial tissue is classified as *simple.* Tissue with more than one layer is called *stratified.* Epithelial tissues are also classified according to shape:

» **Squamous:** *Squamous* cells are thin, flat, and odd-shaped.

» **Cuboidal:** As the name implies, *cuboidal* cells are equal in height and width; shaped like cubes.

» **Columnar:** *Columnar* cells are taller than they are wide.

Following are the ten primary types of epithelial tissues:

» **Simple squamous epithelium:** This flat layer of scale-like cells looks like a fish's scales and is useful in diffusion and filtration. Each cell nucleus is centrally located and is round or oval. Simple squamous epithelium lines the lungs' air sacs where oxygen and carbon dioxide are exchanged.

» **Simple cuboidal epithelium:** These cube-shaped cells, found in a single layer that looks like a microscopic mattress, have centrally located nuclei that usually are round. They are found in the ovaries, line the tubes of the kidneys, and line some glands such as the thyroid, sweat glands, and salivary glands, as well as the inner surface of the eardrum, known as the *tympanic membrane.* Functions of this type of epithelium are secretion, absorption, and tube formation. The kidney tubules have microvilli that increase the area of adsorption for more efficient diffusion and absorption.

» **Simple columnar epithelium:** These densely packed cells are taller than they are wide, with elongated nuclei located near the base of each cell. Found lining the digestive tract from the stomach to the anal canal, the functions of this type of epithelium are secretion and absorption.

» **Ciliated simple columnar epithelium:** A close cousin to simple columnar epithelium, this type of tissue has hairlike cilia that can move mucus and other substances across the cell. It's found lining the small respiratory tubes.

» **Pseudostratified columnar epithelium:** Pay attention to the prefix *pseudo–* here, which means "false." It may look multilayered because the cells' nuclei are scattered at different levels, but it's not. This type of epithelium is found in the large ducts of the parotid glands (salivary glands) and some segments of the male reproductive system, including the urethra.

» **Ciliated pseudostratified columnar epithelium:** Another variation on a theme, this tissue is nearly identical to pseudostratified columnar epithelium. The difference is that the cells on this tissue's free surface have cilia, making it ideal for lining air passages because the cilia's uniform waving action causes dust and dirt particles trapped in a thin layer of mucus to move in one direction — away from the lungs and toward the throat and mouth. It's the insertion of these mucus-producing cells, called goblet cells, that cause the neighboring cells to lose their uniform, columnar shape.

WARNING

>> **Stratified squamous epithelium:** This tissue is the stuff you see every day — your outer skin, or epidermis. Found in areas where the outer cell layer is constantly worn away, this type of epithelium regenerates its surface layer with cells from lower layers.

Though the newly produced cells along the basement membrane look more cuboidal, the cells in the other layers have the characteristic squamous shape. Don't fall into the trap of calling it stratified cuboidal!

>> **Stratified cuboidal epithelium:** This epithelium consists of two or three layers of cube shaped cells. It can be found in sweat glands and the male urethra. Its function is primarily protection of ducts in their neighboring tissue.

>> **Stratified columnar epithelium:** Also multilayered, this epithelium is found lining parts of the male urethra, excretory ducts of glands, the pharynx, and the conjunctiva of the eye. It has a protective function as well as secretion.

>> **Transitional epithelium:** This multilayered epithelium is referred to as *transitional* because its cells can shape-shift from cubes to squamous-like flat surfaces and back again. Found lining the urinary bladder, the cells stretch and flatten out (appearing squamous) to make room for urine. When urination occurs, the cells relax and assume their original form (appearing cuboidal).

Following are some practice questions dealing with epithelial tissue.

EXAMPLE

Q. Stratified epithelial tissue can be described as

a. a thin sheet of cells.

b. covered in cilia.

c. layers of stacked epithelial cells.

d. a long string of tissue.

A. The correct answer is layers of stacked epithelial cells. Remember that "stratified" means layers.

 1–3 Match the epithelial tissue with its location in the body.

a. desmosome

b. gap junction

c. tight junction

1 _____ Binds cells tightly, leaving little to no room between them

2 _____ Tightly connects cells with a tunnel-like protein, allowing them to share their cytoplasmic contents

3 _____ Adheres cells together by linking filaments of their cytoskeleton, allowing the tissue to better resist forces

 4 Transitional epithelial tissue

a. primarily protects specific areas.

b. has the ability to stretch.

c. is constantly worn away and replaced.

d. is tailor-made for absorption.

e. is found in the salivary glands.

5-9 Match the epithelial tissue with its location in the body.

a. Urinary bladder

b. Tubules of the kidney

c. Digestive tract

d. Epidermis of the skin

e. Respiratory passages

5 _____ Simple columnar

6 _____ Stratified squamous

7 _____ Transitional

8 _____ Pseudostratified ciliated columnar

9 _____ Simple cuboidal

10 The key functions of epithelial tissue do **NOT** include

a. protection.

b. secretion.

c. contraction.

d. filtration.

e. absorption.

11-18 Identify the epithelial tissues shown in Figure 5-1.

11. _____

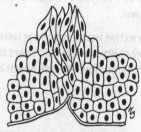

15. _____

12. _____

16. _____

13. _____

17. _____

FIGURE 5-1:
Epithelial
tissues.

14. _____

18. _____

Illustration by Kathryn Born, MA

Making a Connection: Connective Tissue

Connective tissues connect, support, and bind body structures together. Unlike other types of tissues, connective tissues are classified more by the stuff in which the cells lie — the extracellular *matrix* — than by the cells themselves. The matrix contains numerous fibrous proteins embedded in the *ground substance*, which is a thin gel. (It's often referred to as the extracellular fluid.) In most cases, the cells that produce that matrix are scattered within it. The load-bearing strength of connective tissue comes from a fibrous protein called *collagen*. All connective tissues contain a varying mix of collagen, elastic fibers for stretch, and reticular fibers which form a supportive mesh.

REMEMBER

Following are the primary types of connective tissue:

>> **Areolar, or loose, tissue:** This tissue exists between and around almost everything in the body to bind structures together and fill space. It's made up of thick, wavy ribbons of *collagenous fibers,* and cylindrical threads called *elastic fibers,* which contain the protein elastin. Various cells, including lymphocytes, fibroblasts, fat cells, and mast cells, are scattered throughout the ground substance, leaving ample room for blood and lymphatic vessels as well as nerves (see Figure 5-2).

>> **Dense regular connective tissue:** Made up of parallel, densely packed bands or sheets of fibers (see Figure 5-2), this type of tissue is found in tendons, ligaments, surrounding joints, and anchoring organs.

Dense regular connective tissue has great tensile strength that resists lengthwise pulling forces. Ligaments contain more elastic fibers than tendons, making them more flexible. Neither have much blood flow though, which is why an injury to a tendon or ligament takes so long to heal.

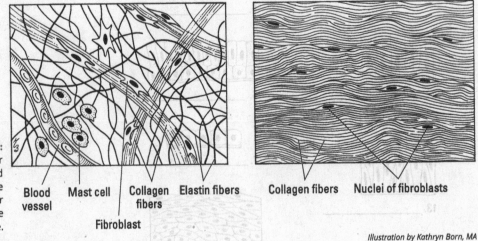

FIGURE 5-2: Areolar tissue and dense regular connective tissue.

Blood vessel   Mast cell   Collagen fibers   Elastin fibers   Collagen fibers   Nuclei of fibroblasts

Fibroblast

Illustration by Kathryn Born, MA

» Dense irregular connective tissue: This tissue has the same components as dense regular but appears much less organized. Its fibers twist and weave around each other, forming a thick tissue that can withstand stresses applied from any direction and leave space for other structures. This tissue makes up the strong inner skin layer called the *dermis* as well as the outer capsule of organs like the kidney and the spleen.

» Adipose tissue: Composed of fat cells, or *adipocytes,* this tissue forms padding around internal organs, provides insulation, and stores energy in fat molecules called triglycerides. Fat molecules fill the cells, forcing the nuclei against the cell membranes and giving them a bubble-like shape. Adipose has an *intracellular matrix* rather than an extracellular matrix.

» Reticular tissue: Literally translated as "web-like" or "net-like," reticular tissue is made up of slender, branching reticular fibers with reticular cells (a specialized fibroblast) overlaying them. Its intricate structure makes it a particularly good filter, which explains why it's found inside the spleen, lymph nodes, and bone marrow.

» Cartilage: These firm but flexible tissues, made up of collagen and elastic fibers, have no nerve cells or blood vessels (a state called *avascular*). Cartilage contains openings called *lacunae* (from the Latin word *lacus* for "lake" or "pit") that enclose mature cells called *chondrocytes.* The ground substance is full of proteoglycans making it thicker than other connective tissues. There are three types of cartilage:

- **Hyaline cartilage:** The most abundant cartilage in the body, it's collagenous and made up of a uniform matrix pocked with chondrocytes. It lays the foundation for the embryonic skeleton, forms the costal (rib) cartilages, makes up nose cartilage, and covers the articulating surfaces of bones that form movable joints.

- **Fibrocartilage:** As the name implies, fibrocartilage contains thick, compact collagen fibers. The sponge-like structure, with the lacunae and chondrocytes lined up within the fibers, makes it a good shock absorber. It's found in the intervertebral discs of the vertebral column and in the symphysis pubis at the front of the pelvis.

- **Elastic cartilage:** Similar to hyaline cartilage but with a much greater abundance of elastic fibers, elastic cartilage has more tightly packed lacunae and chondrocytes between parallel elastic fibers. This cartilage — which makes up structures where a specific form is important such as the outer ear and epiglottis — tends to bounce back to its original shape after being bent.

» Osseus, or bone, tissue: Essentially, bone is mineralized connective tissue, making it the most rigid of the group. The characteristic collagen fibers are coated in mineral deposits giving it the hardness we associate with our bones. Cells called *osteoblasts* mineralize the matrix into repeating patterns called *osteons.* In the center of each osteon is a large opening, the *Haversian canal* (or *central canal*) that contains blood vessels, lymph vessels, and nerves. The central canal is surrounded by thin layers of bony matrix called *lamellae* that contain the lacunae, which house *osteocytes* (mature bone cells). Smaller tunnels called *canaliculi* connect the lacunae and the arms of the osteocytes form gap junctions within them. (We explore bone in more detail in Chapter 7.)

» Blood: Yes, blood is considered a type of connective tissue. Like other connective tissues, it has an extracellular matrix — in this case, plasma — in which *erythrocytes* (red blood cells), *leukocytes* (white blood cells), and *thrombocytes* (platelets) are suspended. (Blood also is considered a vascular tissue because it circulates inside arteries and veins, and we get into that in Chapter 11.) Roughly half of blood's volume is fluid or plasma while the other half is suspended cells. Erythrocytes are concave on both sides and contain a pigment, *hemoglobin,* which carries oxygen throughout the body and takes carbon dioxide away. There are approximately 5 million erythrocytes per cubic millimeter of whole blood. Thrombocytes, which number approximately 250,000 per cubic millimeter, are fragments of cells used in blood clotting. Some leukocytes are large *phagocytic* cells (literally "cells that eat") that are part of the body's immune system. There are, however, relatively few of them — less than 10,000 per cubic millimeter.

 19 What is adipose tissue composed of?

a. Mast cells

b. Chondrocytes

c. Osteocytes

d. Leukocytes

e. Fat cells

 20 What are tendons composed of?

a. Elastic cartilage

b. Dense regular connective tissue

c. Areolar tissue

d. Fibrocartilage

e. Dense irregular connective tissue

21 Hyaline cartilage

a. covers the surface of articulating bones.

b. lines the symphysis pubis.

c. forms the epiglottis.

d. forms the outer ear and nose.

e. cushions intervertebral discs.

 22 Why does dense irregular connective tissue twist and weave around?

a. To fill gaps between other tissues

b. To withstand stresses applied from any direction

c. To prevent dead or dying cells from weakening its structure

d. To provide stretch and elasticity

e. To provide firm connections between muscles and bones

23 Which tissue contains lacunae with osteocytes?

a. Elastic cartilage

b. Bone

c. Hyaline cartilage

d. Blood

e. Reticular

24 What function do thrombocytes serve?

a. They block invading microbes.

b. They carry oxygen to connective tissues.

c. They nourish osteocytes.

d. They contribute to blood clotting.

e. They provide nutrients to erythrocytes.

Flexing It: Muscle Tissue

Although we review how muscles work in Chapter 8, in histology you should know that muscle tissue is made up of cells, which are referred to as fibers, known as *myocytes*. The cytoplasm within the fibers is called *sarcoplasm*, and within that sarcoplasm are minute *myofibrils* that contain the protein filaments *actin* and *myosin*. These filaments slide past each other during a muscle contraction, shortening the fiber.

REMEMBER

Following are the three types of muscle tissue (see Figure 5-3):

» **Skeletal muscle tissue:** Biceps, triceps, pecs — these are the muscles that bodybuilders focus on. As the name implies, skeletal muscles attach to the skeleton and are controlled throughout by the conscious (voluntary) control function of the central nervous system for movement. Muscle fibers are cylindrical with several nuclei in each cell (which makes them *multinucleated*) and obvious *striations* (stripes that span the width of the cell) throughout. They are bundled in parallel and are roughly the length of the muscle they make up.

» **Cardiac muscle tissue:** This tissue is composed of cylindrical fibers, *cardiomyocytes*, each with a central nucleus and are striated. The tissue forms a branching pathway as one cell can join with two or more fibers at one end using gap junctions called *intercalated discs*. Cardiac muscle tissue contractions occur through the autonomic nervous system (involuntary control).

» **Smooth muscle tissue:** Made up of spindle-shaped fibers with large, centrally located nuclei, it's found in the walls of internal organs, or *viscera*. As with cardiac muscle, this type of tissue contracts without conscious control. Smooth muscle gets its name from the fact that, unlike other muscle tissue types, it is not striated.

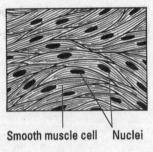 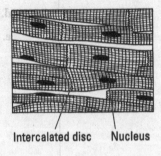 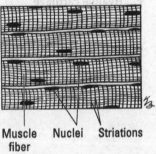

FIGURE 5-3:
Muscle tissues: Smooth, cardiac, and skeletal.

Smooth muscle cell   Nuclei   Intercalated disc   Nucleus   Muscle fiber   Nuclei   Striations

Illustration by Kathryn Born, MA

---

25 Which type of tissue is multinucleated?

a. Skeletal muscle tissue

b. Cardiac muscle tissue

c. Smooth muscle tissue

d. Autonomic muscle tissue

26 Both skeletal and cardiac muscle tissue have prominent lines across the fiber. What are they called?

a. Fibroblasts

b. Multinucleates

c. Lacunae

d. Striations

e. Intercalated discs

27 Which of the following contains smooth muscle tissue?

a. The heart

b. The urinary bladder

c. The biceps

d. The deltoid

28 Describe the location and function of intercalated discs.

_____

_____

_____

_____

_____

Getting the Signal Across: Nervous Tissue

There's only one type of nervous tissue and only one primary type of cell in it: the *neuron*. There are numerous other cells, called *glial cells*, but they don't perform the communication function we associate with nervous tissue. A neuron is unique in that it can both generate and conduct electrical signals in the body. That process starts when sense receptors receive a stimulus that causes an electrical signal to be sent through cytoplasmic projections called *dendrites*. From there, the signal moves through the *cell body* to the *axon*, another cytoplasmic extension. Here, an impulse is generated and spreads to the end of the axon, where it hands the signal off chemically to the next cell down the line. (We look more closely at how all that happens in Chapter 9.)

 29 True or False: Nervous tissue is essentially formed by neurons connecting end-to-end with very little matrix.

 30 Which statement best describes how neurons communicate?

a. Axons generate impulse, which is passed to dendrites through the cell body.

b. Axons receive a stimulus, which is passed to the cell body, triggering the dendrites to release a chemical.

c. Cell bodies respond to a stimulus by generating an electrical signal to pass on to the dendrites and axons, where the former will pass it on electrically and the latter will do so chemically.

d. In response to stimulus, the dendrite generates an electrical signal that is passed to the cell body. The axon will then spread an impulse down its length, releasing a chemical at the end.

e. Dendrites respond to stimulus by releasing a chemical within the cell that travels through the cell body. The axon generates an impulse in response, spreading it to the end, where it will pass the signal on to the next cell.

Answers to Questions on Histology

The following are answers to the practice questions presented in this chapter.

(1-3) 1. **c. tight junction**; 2. **b. gap junction**; 3. **a. desmosome**

(4) Transitional epithelial tissue **b. has the ability to stretch.** This tissue lines the bladder, so it had better be stretchy!

(5) Simple columnar: **c. Digestive tract**

(6) Stratified squamous: **d. Epidermis of the skin**

(7) Transitional: **a. Urinary bladder**

(8) Pseudostratified ciliated columnar: **e. Respiratory passages**

(9) Simple cuboidal: **b. Tubules of the kidney**

(10) The key functions of epithelial tissue do *not* include **c. contraction.** Muscles contract; epithelial tissue does not.

(11-18) Following is how Figure 5-1, the types of epithelial cells and tissues, should be labeled.

11. **stratified cuboidal;** 12. **simple columnar** 13. **pseudostratified;** 14. **simple squamous;** 15. **transitional;** 16. **stratified columnar;** 17. **simple cuboidal;** 18. **stratified squamous**

WARNING

Identifying epithelials can be a bit tricky. First look for the basement membrane and the space then look only at the cells in between. Simple cuboidal can appear stratified because the tissue forms tubes and neighboring tubes will connect at their basement membrane. When stratified, you must look at the pattern of the cells throughout. Stratified squamous has cube-shaped cells along the bottom but the bulk of the cells are clearly squamous. Transitional is the most troublesome but you will see both squamous and cuboidal throughout the width of the tissue.

(19) Adipose tissue is composed of **e. fat cells.** Think of the Latin *adeps*, which means "fat."

(20) Tendons are composed of **b. dense regular connective tissue.** Tendons are dense and exhibit a regular pattern.

(21) Hyaline cartilage **a. covers the surface of articulating bones.** It's all in the names of the three types of cartilage. Hyaline comes from the Greek word for *glass*, which seems appropriate for something covering a surface. Fibrocartilage contains thick bunches of fibers, which doesn't sound right for covering a surface. And elastic cartilage tends to be found where exact shapes are important, such as in the outer ear.

(22) Why does dense irregular connective tissue twist and weave around? **b. To withstand stresses applied from any direction.** The irregular pattern also creates space for structure like blood vessels to pass through.

(23) Which tissue contains lacunae with osteocytes? **b. Bone.** Remember that *osteo* is the Latin word for "bone."

(24) What function do thrombocytes serve? **d. They contribute to blood clotting.** Knowing that the Greek word *thrombos* means "clot" can help you spot the correct answer in this question.

(25) Which type of tissue is multinucleated? **a. Skeletal muscle tissue.** Cardiac muscle tissue and smooth muscle tissue typically have only one nucleus per cell; autonomic muscle tissue is a trick answer choice.

(26) Both skeletal and cardiac muscle tissue have prominent lines across the fiber. What are they called? **d. Striations.** Striations — think *stripes* — are alternating light and dark lines.

(27) Which of the following contains smooth muscle tissue? **b. The urinary bladder.** The other answer choices contain striated tissue, which technically means that they aren't smooth.

(28) Describe the location and function of intercalated discs. **Intercalated discs are found where cardiac muscle fibers meet end-to-end. These gap junctions form tunnels that allow cells to pass molecules to each other directly, without expending energy on endo- and exocytosis.**

(29) Nervous tissue is essentially formed by neurons connecting end-to-end with very little matrix. **False:** Abundant space is visible between neurons dotted with glial cells and numerous "fibers," which are actually the ends of axons belonging to other neurons.

(30) Which statement best describes how neurons communicate? **d. In response to stimulus, the dendrite generates an electrical signal that is passed to the cell body. The axon will then spread an impulse down its length, releasing a chemical at the end.** While neurons communicate with themselves using electrical signaling, that cannot be passed to another cell so chemicals must be used. While both dendrites and axons can generate an electric signal, it is not termed impulse until the axon does it.

Weaving It Together: Bones, Muscles, and Skin

2

IN THIS PART . . .

Wrap the package in the body's largest single organ: the skin. Get the scoop on the epidermis and the dermis; find out about nerves in the skin; and accessorize with hair, nails, and glands.

Build on a strong foundation: Explore how bones are formed. Piece those bones together into a skeleton and distinguish your axial from your appendicular.

Articulate some joints and attach muscles to that framework. Contract and relax to leverage muscle power.

Chapter 6

It's Skin Deep: The Integumentary System

id you know that the skin is the body's largest organ? In an average person, its 17 to 20 square feet of surface area represents 5-7 percent of the body's weight. Self-repairing and surprisingly durable, the skin is the first line of defense against the harmful effects of the outside world. It helps retain moisture; regulates body temperature; hosts the sense receptors for touch, pain, and heat; and excretes excess salts and small amounts of waste.

Skin is jam-packed with components; it has been estimated that every square inch of skin contains 15 feet of blood vessels, 4 yards of nerves, 650 sweat glands, 100 oil glands, 1,500 sensory receptors, and more than 3 million cells with an average life span of 26 days that are constantly being replaced.

In this chapter, we peel back the surface of this most-visible organ system. We also give you plenty of opportunities to test your knowledge.

Digging Deep into Dermatology

Skin — together with hair, nails, and glands — composes the *integumentary system* (shown in Figure 6-1). The name stems from the Latin verb *integere*, which means "to cover." The relevant Greek and Latin roots include *dermato* and *cutis*, both of which mean "skin."

The skin consists of two primary parts, which we describe in the following sections: the *epidermis* and the *dermis*. (The Greek root *epi*– means "upon" or "above.") Underlying the epidermis and dermis is the *hypodermis* or *superficial fascia* (also sometimes called *subcutaneous tissue*), which acts as a foundation but is not technically part of the skin. Composed of areolar and adipose tissue, it anchors the skin through fibers that extend from the dermis. Underneath, it attaches loosely to tissues and organs so that muscles can move freely. The fat in the hypodermis buffers deeper tissues and acts as insulation, preventing heat loss from within the body's core. The dermis and the hypodermis are home to pressure-sensitive nerve endings called *lamellated* or *Pacinian corpuscles* that respond to a deeper poke in the skin.

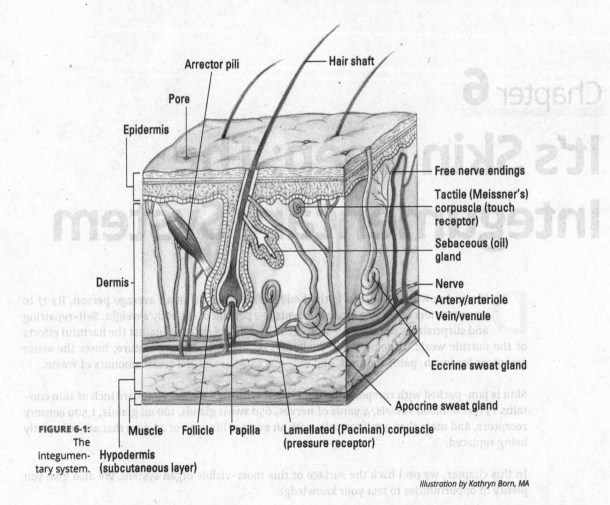

FIGURE 6-1: The integumentary system.

Illustration by Kathryn Born, MA

The epidermis: Don't judge this book by its cover

REMEMBER

The epidermis, which contains no blood vessels, is made up of layers of closely packed epithelial cells. From the outside in, these layers are the following:

>> **Stratum corneum** (literally the "horny layer") is about 20 layers of flat, scaly, dead cells containing a type of water-repellent protein called *keratin*. These cells, which represent about three-quarters of the thickness of the epidermis, are said to be *cornified,* which means that

they're tough and horny like the cells that form hair or fingernails. Humans shed this layer of tough, durable skin at a prodigious rate. Where the skin is rubbed or pressed more often, cell division increases, resulting in calluses and corns.

>> **Stratum lucidum** (from the Latin word for "clear") is found only in the thick skin on the palms of the hands and the soles of the feet. This translucent layer of dead cells contains *eleidin,* a protein that becomes keratin as the cells migrate into the stratum corneum, and it consists of cells that have lost their nuclei and cytoplasm, hence they appear clear.

>> **Stratum granulosum** is three to five layers of flattened cells containing *keratohyalin,* a substance that marks the beginning of keratin formation. No nourishment from blood vessels reaches this far into the epidermis, so cells are either dead or dying by the time they reach the stratum granulosum. The nuclei of cells found in this layer are degenerating; when the nuclei break down entirely, the cell can't metabolize nutrients and dies.

>> **Stratum spinosum** (also sometimes called the *spinous layer*) has five to ten layers containing *spinous,* or *prickle cells,* named for the spinelike projections that connect them with other cells in the layer. *Langerhans cells,* which are involved in the body's immune response, are prevalent in the upper portions of this layer and sometimes the lower part of the stratum granulosum; they migrate from the skin to the lymph nodes in response to infection.

>> **Stratum basale** (or *stratum germinativum*) is also referred to as the *germinal layer* because this single layer is mostly composed of cube-shaped stem cells. This is the site of mitosis as these stem cells divide and push cells upward. It takes about two weeks for the cells that originate here to migrate up to the stratum corneum, and it's another two weeks before they're shed. It rests on the papillary (rough or bumpy) surface of the dermis, close to the blood supply needed for nourishment and oxygen. About a quarter of this layer's cells are *melanocytes,* cells that synthesize a pale yellow to black pigment called *melanin* that contributes to skin color and provides protection against ultraviolet radiation (the kind of radiation found in sunlight). The remaining basal cells divide into *keratinocytes,* the primary epithelial cell of the skin. Melanocytes secrete melanin and the keratinocytes bring it in via a process called *cytocrine secretion,* thus pigmenting the cells throughout the epithelium. *Merkel cells,* large nervous cells involved in the sense of touch, occasionally appear amid the keratinocytes.

In addition to melanin, the epidermis contains a yellowish pigment called *carotene* (the same one found in carrots and sweet potatoes). Found in the stratum corneum and the fatty layers beneath the skin, it produces the yellowish hue associated with Asian ancestry or increased carrot consumption. The pink to red color of Caucasian skin is caused by *hemoglobin,* the red pigment of the blood cells. Because Caucasian skin contains relatively less melanin, hemoglobin can be seen more easily through the epidermis. Sometimes the limited melanin in Caucasian skin pools in small patches. Can you guess the name of those patches of color? Yep, they're freckles. Albinos, on the other hand, have no melanin in their skin at all, making them particularly sensitive to ultraviolet radiation.

Deep to the stratum basale is the *basement membrane* of the epidermal tissue. It is thought to be created by the layer itself. In addition to anchoring the epidermal layer to the dermis below. It also acts as a filter — controlling what gets up into this layer from below and what is allowed deeper.

Uplifted portions of the dermis, *dermal papillae* (fingerlike projections), cause ridges and grooves to form on the outer surface of the epidermis that increase the friction needed to grasp objects or move across slick surfaces. On hands and feet, these ridges form patterns of loops and whorls — fingerprints, palm prints, and footprints — that are unique to each person. You leave these imprints on smooth surfaces because of the oily secretions of the oil and sweat glands on the skin's surface.

The dermis: Going more than skin deep

REMEMBER

Beneath the epidermis is a thicker, fibrous structure called the dermis. It consists of the following two layers, which blend together:

>> **The outer, soft *papillary layer*** contains areolar tissue that projects into the epidermis to bring blood and nerve endings closer. The papillae increase the surface area of the connection between the dermis and epidermis' basement membrane. Some of these papillae contain *Meissner's corpuscles,* nervous receptors that are sensitive to soft touch.

>> **The inner, thicker *reticular layer*** (from the Latin word *rete* for "net") is made up of dense irregular connective tissue containing interlacing bundles of collagenous and elastic fibers that form the strong, resistant layer used to make leather and suede from animal hides. This layer is what gives skin its strength, extensibility, and elasticity. Within the reticular layer are *sebaceous glands* (oil glands), sweat glands, fat cells, hair follicles, nerve cells, and larger blood vessels.

The dermis also includes *fibroblasts* (which create the fibers that form the bulk of connective tissues), and *macrophages* (which engulf waste and foreign microorganisms). Thickest on the palms of the hands and soles of the feet, the dermis is thinnest over the eyelids, penis, and scrotum. It's thicker on the back (posterior) than on the stomach (anterior) and thicker on the sides (lateral) than toward the middle of the body (medial). The various skin "accessories" — blood vessels, nerves, glands, and hair follicles — are embedded here. Adipocytes (fat cells) are found at the bottom of the dermis, which transitions into the hypodermis below.

See if you've got the skinny on skin so far:

EXAMPLE

Q. The layer of epidermis that's composed of a horny, cornified tissue that sloughs off is called the stratum

a. corneum.

b. lucidum.

c. granulosum.

d. spinosum.

e. basale.

A. The correct answer is corneum. Cornified, corns — think of how hard a kernel of popcorn can be.

1. What does the papillary layer of the dermis *not* do?

   a. Filter out microbes

   b. Extend into the epidermis

   c. Carry the blood and nerve endings close to the epidermis

   d. Aid in holding the epidermis and dermis together

   e. Contain cells sensitive to touch

2. The epidermal ridges on the fingers function to

   a. provide a means of identification.

   b. increase the friction of the epidermal surface.

   c. decrease water loss by the tissues.

   d. aid in regulating body temperature.

   e. prevent bacterial infection.

3. Caucasian skin color is caused by

   a. carotene pigment in the dermis.

   b. the high level of melanin in the epidermis.

   c. less melanin in the skin, allowing the blood pigment to be seen.

   d. the absence of all pigment.

   e. melanin and carotene pigments.

4. The sequence of layers in the epidermis from the dermis outward is

   a. corneum, lucidum, granulosum, spinosum, basale.

   b. corneum, granulosum, lucidum, basale, spinosum.

   c. spinosum, basale, granulosum, corneum, lucidum.

   d. basale, spinosum, granulosum, lucidum, corneum.

   e. basale, lucidum, corneum, spinosum, granulosum.

5 What's another name for the subcutaneous layer of tissue?

   a. Epidermis
   b. Superficial fascia
   c. Papillary layer
   d. Inner reticular layer
   e. Dermis

6 What do carrots and sweet potatoes have in common with human epidermis?

   a. They synthesize melanocytes.
   b. They have layers of cells at varying stages of development.
   c. They contain the pigment carotene.
   d. They secrete water proofing oils.
   e. They have a papillary surface.

7 What is the name of a layer of dense irregular connective tissue containing interlacing bundles of collagenous and elastic fibers?

   a. Basale layer of the epidermis
   b. Reticular layer of the dermis
   c. Outer layer of the hypodermis
   d. Papillary layer of the dermis
   e. Inner layer of the hypodermis

8 Why is keratin important to the skin?

   a. It makes the stratum corneum thick, tough, and water-repellant.
   b. It keeps the stratum lucidum moisturized.
   c. It helps nourish the stratum granulosum.
   d. It maintains connections between the cells of the stratum spinosum.
   e. It distributes sweat evenly through the epidermal layer.

9 The stratum _____ epidermal layer contains keratohyalin.

a. germinativum

b. spinosum

c. lucidum

d. granulosum

e. corneum

10 What do the prickle cells in the spinous layer do?

a. They metabolize nutrients.

b. They connect the cells of the layer.

c. They transform other cells that have lost their nuclei and cytoplasm.

d. They move nutrients from the blood into the epidermis.

e. They secrete keratohyalin.

11 What is the name for melanin that forms into patches?

a. Flexion creases

b. Freckles

c. Dermal papillae

d. Lamellated corpuscles

e. Matrix

12 Briefly outline the steps of epidermal growth.

_____

_____

_____

_____

_____

_____

Touching a Nerve in the Integumentary System

At least four kinds of receptors are involved in creating the sensation of touch:

>> **Free nerve endings:** These are the most abundant type of sensory endings, occurring widely in the integument and within muscles, joints, viscera, and other structures. Nerve endings are *dendrites* (branched extensions) of sensory neurons with a swelling at the end. These mostly function in *nociception*, or pain sensation, though some sense temperature, touch, and stretch. Found all over the body, free nerve endings are especially prevalent in epithelial and connective tissue.

The free nerve endings that sense temperature come in two varieties: ones that respond to cold stimuli and different ones that respond to heat. A warm sensation comes from the hypothalamus of the brain interpreting information from both.

>> **Meissner's corpuscles:** These light-touch mechanoreceptors lie within the dermal papillae (you can see one in Figure 6-1). They're small, egg-shaped capsules of connective tissue surrounding a spiraled end of a dendrite. They are the most abundant in sensitive skin areas such as the lips and fingertips.

>> **Pacinian corpuscles:** These deep-pressure mechanoreceptors are dendrites surrounded by concentric layers of connective tissue (check one out in Figure 6-1). Found deep within the dermis and hypodermis, they respond to deep or firm pressure and vibrations. Each is more than 2 millimeters long and therefore visible to the naked eye.

>> **Hair plexus:** These nerve endings wrap around the hair follicle and function as mechanoreceptors. When the hair is moved, it stimulates the nerve endings, creating a very light touch sensation.

Test whether you're staying in touch with this section:

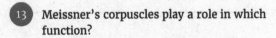

13 Meissner's corpuscles play a role in which function?

   a. The sensation of heavy pressure

   b. The skin's ability to rebound into shape after pressure is applied

   c. The immediate withdrawal response from intense heat

   d. The sense of motion

   e. The sensation of soft touch

14 Which receptor type responds to a painful stimulus?

   a. Hair plexus

   b. Meissner's corpuscle

   c. Pacinian corpuscle

   d. Thermoreceptor

   e. Nociceptor

Accessorizing with Hair, Nails, and Glands

Mother Nature has accessorized your fashionable over-wrap with a variety of specialized structures that grow from the epidermis: hair, fingernails, toenails, and several types of glands.

Wigging out about hair

Like most mammals, hair covers the entire human body except for the lips, eyelids, palms of the hands, soles of the feet, nipples, and portions of external reproductive organs. But human body hair generally is sparser and much lighter in color than that sported by most other mammals. Animals have hair for protection and temperature control. For humans, however, body hair is largely a secondary sex characteristic.

A thick head of hair protects the scalp from exposure to the sun's harmful rays and limits heat loss. Eyelashes block sunlight and deflect debris from the eyes. Hair in the nose and ears prevents airborne particles and insects from entering. Touch receptors connected to hair follicles respond to the lightest brush.

The average adult has about 5 million hairs, with about 100,000 of those growing from the scalp. Normal hair loss from an adult scalp is anywhere from 40 to 100 hairs each day, although baldness can result from genetic factors, hormonal imbalances, scalp injuries, disease, dietary deficiencies, radiation, or chemotherapy.

Each hair grows at an angle from a follicle embedded in the epidermis and extending into the dermis (refer to Figure 6-1); scalp hairs sometimes reach as far as the hypodermis. Nerves reach the hair at the follicle's expanded base, called the *bulb*, where a nipple-shaped papilla of connective tissue and capillaries provide nutrients to the growing hair. Epithelial cells in the bulb divide to produce the hair's *root*. Like the growth of epidermis, cells die as they are pushed away from the source of nutrients. By the time the hair reaches the scalp's surface, now referred to as the *shaft*, the cells have long been dead. The shape of a hair's cross section can vary from round to oval or even flat — influenced by the follicle's shape. Oval hairs grow out appearing wavy or curly, flat hairs appear kinky, and round hairs grow out straight. Each scalp hair grows for two to three years at a rate of about 1/3 to 1/2 millimeter per day, or 10 to 18 centimeters per year. When mature, the hair rests for three or four months before slowly losing its attachment. Eventually, it falls out and is replaced by a new hair.

Hair pigment (which is mostly melanin, just as in the skin) is produced by melanocytes in the follicle and transferred to the cells. Three types of melanin — black, brown, and yellow — combine in different quantities for each individual to produce different hair colors ranging from light blonde to black. Gray and white hairs grow in when melanin levels decrease and air pockets form where the pigment used to be.

Wondering why you have to shampoo so often? Hair becomes oily over time thanks to *sebum*, a mixture of cholesterol, fats, and other substances secreted from a *sebaceous* (or *holocrine*) gland found next to each follicle (see one for yourself in Figure 6-1). Sebum keeps both hair and skin soft, pliable, and waterproof. Attached to each follicle is a smooth muscle called an *arrector pili* (literally "raised hair") that straightens the hair shaft, elevating the skin in a pattern called *goose bumps* or *goose pimples.* In other mammals this traps heat, keeping them warm. We, however, do not have enough hair for that benefit, but we certainly still get them!

The oil slick

The dermis is also home to our *sebaceous glands*, which are most often associated with a hair follicle (see Figure 6-1). They secrete a fatty substance called *sebum* which you know as oil. Since the sebaceous gland is attached to the follicle, when the arrector pili contract, the pressure put on the gland can cause it to release its sebum. While we find the accumulation of this oil to be annoying, especially since it's the main cause of acne, it does have important functions. It softens the hairs and skin surface, making them both more pliable. It also serves to reduce friction as well as provide waterproofing.

Nailing the fingers and toes

The hard part of the fingernails and toenails is also made of keratin. Human nails (which actually are vestigial claws) have three parts: a *root bed* at the nail base, a *body* that's attached to the fingertip, and a *free edge* that grows beyond the end of the finger or toe.

Heavily cornified tissue forms the nails from modified strata corneum and lucidum (we describe these and other layers in the earlier section "The epidermis: Don't judge this book by its cover"). *Cornification* is the process by which squamous cells are basically turned into keratin. A narrow fold of the stratum corneum turns back to form the *eponychium*, or *cuticle*. Under the nail, the nail bed is formed by the strata basale and spinosum. At the base of the nail, partially tucked under the cuticle, the strata thicken to form a whitish area called the *lunula* (literally "little moon") that can be seen through the nail. Beneath the lunula is the *nail matrix*, a region of thickened strata where mitosis pushes previously formed cornified cells forward, making the nail grow. Under the free edge of the nail, the stratum corneum thickens to form the *hyponychium*. Nails are pinkish in color because of hemoglobin in the blood of the underlying capillaries, which are visible through the translucent cells of the nail.

On average, fingernails grow about 1 millimeter each week. Toenails tend to grow even more slowly. Nails function as an aid to grasping, as a tool for manipulating small objects, and as protection against trauma to the ends of fingers and toes.

Sweating the details

Humans perspire over nearly every inch of skin, but anyone with sweaty palms or smelly feet can attest to the fact that sweat glands are most numerous in the palms and soles, with the forehead running a close third. There are two types of sweat, or *sudoriferous*, glands: eccrine and apocrine. Both are coiled tubules embedded in the dermis and are composed of simple columnar cells (refer to Figure 6-1).

>> **Eccrine sweat glands** are distributed widely over the body — an average adult has roughly 3 million of them — and produce the watery, salty secretion you know as sweat. Each gland's duct passes through the epidermis to the skin's surface, where it opens as a *sweat pore*. The sympathetic division of the autonomic nervous system (see Chapter 9) controls when and how much perspiration is secreted depending on how hot the body becomes. Sweat helps cool the skin's surface by evaporating as fast as it forms. About 99 percent of eccrine-type sweat is water, but the remaining 1 percent is a mixture of sodium chloride and other salts, uric acid, urea, amino acids, ammonia, sugar, lactic acid, and ascorbic acid.

>> **Apocrine sweat glands** are located primarily in armpits (known as the *axillary region*) and the external genital area. Rather than secreting onto the surface of the skin, these release sweat into the hair follicle. These glands don't become active until puberty and are triggered by stress. Although apocrine-type sweat contains the same basic components as eccrine sweat and is also odorless when first secreted, bacteria residing in the hair follicle quickly begin to break down its additional fatty acids and proteins — explaining the post-exercise underarm stench. In addition to exercise, sexual and other emotional stimuli can cause contraction of cells around these glands, releasing sweat.

Getting an earful

The occasionally troublesome yellowish substance known as earwax is secreted in the outer part of the ear canal from modified sudoriferous glands called *ceruminous glands* (the Latin word *cera* means "wax"). Lying within the subcutaneous layer of the ear canal, these glands have ducts that either open directly into the ear canal or empty into the ducts of nearby sebaceous glands, mixing their secretion with sebum and dead epithelial cells. Technically called *cerumen*, earwax is the combined secretion of these two glands. Working with ear hairs, cerumen traps any foreign particles before they reach the eardrum. As the cerumen dries, it flakes and falls from the ear, carrying particles out of the ear canal.

Think you've got a grip on everything to do with hair, nails, and glands? Find out by answering the following practice questions:

15　What is another name for the cuticle?

a.　Lunula

b.　Hyponychium

c.　Eponychium

d.　Nail matrix

e.　Perinychium

16　The _____ glands form perspiration.

a.　sebaceous

b.　ceruminous

c.　endocrine

d.　Merkel

e.　sudoriferous

17   The cause of graying hair is
     a.   production of melanin in the shaft of the hair.
     b.   production of carotene in the shaft of the hair.
     c.   decrease in blood supply to the hair.
     d.   lack of melanin in the shaft of the hair.
     e.   parenthood.

18   From where does the hair develop?
     a.   Arrector pili
     b.   Shaft
     c.   Follicle
     d.   Sebaceous gland
     e.   Lanugo

19   The nails are modifications of the epidermal layers
     a.   corneum and lucidum.
     b.   lucidum and granulosum.
     c.   granulosum and spinosum.
     d.   spinosum and basale.
     e.   lucidum and spinosum.

20   Which factor is not associated with baldness?
     a.   Genetics
     b.   Hormonal imbalances
     c.   Scalp injuries
     d.   Lack of melanin
     e.   Disease

21 Sebaceous glands

   a. produce a watery solution called sweat.

   b. produce an oily mixture of cholesterol, fats, and other substances.

   c. produce a waxy secretion called cerumen.

   d. accelerate aging.

   e. are associated with endocrine glands.

22 What function does the bulb at the base of a hair follicle serve?

   a. To prevent dirt and debris from becoming embedded in the skin

   b. To establish additional thermal protection

   c. To provide nutrients to the growing hair

   d. To regulate sweat production

   e. To inject melanin into the hair

23 This gland secretes water with fatty acids and proteins in response to stress. It acquires an unpleasant odor when bacteria breaks down the organic molecules it secretes.

   a. Apocrine sweat gland

   b. Sebaceous gland

   c. Ceruminous gland

   d. Eccrine sweat gland

   e. Mammary gland

24-27 Use the terms that follow to identify the structures of the nail shown in Figure 6-2.

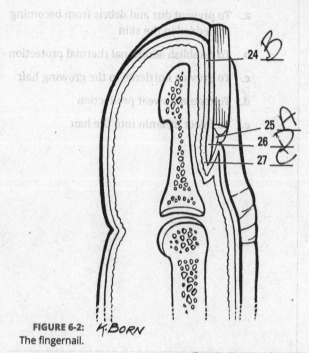

FIGURE 6-2: *K. BORN*
The fingernail.

Illustration by Kathryn Born, MA

a. cuticle

b. nail bed

c. matrix

d. root

---

 28 When the arrector pili contracts, which gland could incidentally release its contents?

a. Apocrine sweat gland

b. Eccrine sweat gland

c. Ceruminous gland

d. Sebaceous gland

e. Endocrine gland

Answers to Questions on the Skin

The following are answers to the practice questions presented in this chapter.

① What does the papillary layer of the dermis not do? **a. Filter out microbes.** Busy little finger-like projections, those papillae, but they're not filters. That function is carried out by the basement membrane of the epidermis.

② The epidermal ridges on the fingers function to **b. increase the friction of the epidermal surface.** They're Mother Nature's way of helping you cling to tree branches or grab food. While Choice a seems reasonable, that's a perk not a function.

③ Caucasian skin color is caused by **c. less melanin in the skin, allowing the blood pigment to be seen.** Here's a fun experiment: Turn off the lights, press your fingers together, and hold a flashlight under them. See the red glow? That's hemoglobin, too.

④ The sequence of layers in the epidermis from the dermis outward is **d. basale, spinosum, granulosum, lucidum, corneum.**

Memory tool time: Base, spine, grain, Lucy, corny. Or try the first letters of Be Super Greedy, Less Caring. Insensitive, yes, but effective.

⑤ What's another name for the subcutaneous layer of tissue? **b. Superficial fascia.** Subcutaneous is the same as hypodermis (from the Greek *hypo–* for "beneath").

⑥ What do carrots and sweet potatoes have in common with human epidermis? **c. They contain the pigment carotene.** Most of the other answers wouldn't apply to both mammalian and vegetable cells.

⑦ What is the name of a layer of dense irregular connective tissue containing interlacing bundles of collagenous and elastic fibers? **b. Reticular layer of the dermis.** The description in this question sounds like a tough structure, so it may help you to remember that the reticular layer is what's used to make leather from animal hides.

⑧ Why is keratin important to the skin? **a. It makes the stratum corneum thick, tough, and water-repellant.** Associate the words "corneum" and "keratin," and you're in great shape.

⑨ The stratum **d. granulosum** epidermal layer contains keratohyalin. Keratohyalin eventually becomes keratin, so think of the layer where the cells are starting to die off.

⑩ What do the prickle cells in the spinous layer do? **b. They connect the cells of the layer.** The spine-like projections that make those connections also make the cells look prickly, hence the name. Don't get distracted by this layer being moisturized. The keratin doesn't do this; the sebum does.

⑪ What is the name for melanin that forms into patches? **b. Freckles.** Ever noticed how kids have more freckles at the end of a long summer spent outdoors? That's ultraviolet radiation working on those melanin patches.

⑫ Outline the steps of epidermal growth:

1 Along the basement membrane, cells in the stratum basale undergo mitosis. As a result, older cells are pushed upward.

2 Cells are pushed into the stratum spinosum, where they spend most of their time creating proteins and fibers for desmosomes, which hold all the cells tightly together. They also take up melanin produced by melanocytes in the basal layer.

3   Since nutrients must pass through the basement membrane, cells receive less and less access to them as they are pushed up. Once they reach the stratum granulosum, they are usually dead. The keratohyalin contained in the cells is the precursor to keratin.

4   By the time the cells reach the stratum corneum, they are unrecognizable. They are completely flattened and mostly just keratin.

The hair and nails are formed by this same general process so you only have to memorize it once!

TIP

13   Meissner's corpuscles play a role in which function? **e. The sensation of soft touch.** Although it's true that several different nerves are involved in the overall sense of touch, Meissner's corpuscles are the most responsive to touch.

14   Which receptor type responds to a painful stimulus? **e. Nociceptor.** Even with extreme heat or pressure, only the nociceptors can send a message that the brain interprets as painful.

15   What is another name for the cuticle? **c. Eponychium.** Recall that the prefix *ep–* refers to "upon" or "around," whereas the prefix *hypo–* refers to "below" or "under." The cuticle is around the base of the nail, so it's the eponychium, not the hyponychium.

16   The **e. sudoriferous** glands form perspiration. The Latin word *sudor* means "sweat."

17   The cause of graying hair is **d. lack of melanin in the shaft of the hair.** Despite the medical cause, people often suspect that answer "e. parenthood" has a lot to do with graying hair.

18   From where does the hair develop? **c. Follicle.** The Latin translation of this word is "small cavity" or "sac," so it makes sense that this would be an origination place.

19   The nails are modifications of the epidermal layers **a. corneum and lucidum.** These are the two upper layers.

20   Which factor is not associated with baldness? **d. Lack of melanin.** This answer just means that your hair won't turn gray, not necessarily that it will fall out of your scalp.

21   Sebaceous glands **b. produce an oily mixture of cholesterol, fats, and other substances.** That secretion's called *sebum* — hence "sebaceous glands."

22   What function does the bulb at the base of a hair follicle serve? **c. To provide nutrients to the growing hair.** This is where connective tissue and capillaries come together to provide those nutrients.

23   This gland contains secretes water with fatty acids and proteins in response to stress. It acquires an unpleasant odor when bacteria breaks down the organic molecules it secretes. **a. Apocrine sweat gland.** These are the truly stinky sweat glands.

Here's a memory tool for the difference between apocrine and eccrine sweat glands: You may have to APOlogize for your APOcrine glands but not your eccrine glands.

TIP

24–27   Following is how Figure 6-2, the fingernail, should be labeled:

24. **b. nail bed**; 25. **a. cuticle**; 26. **d. root**; 27. **c. matrix**

28   When the arrector pili contracts, which gland could incidentally release its contents? **d. Sebaceous gland.** None of the other glands are the closely associated with a hair follicle; an apocrine sweat gland does secrete into the follicle but only the duct is near the follicle.

Chapter 7

A Scaffold to Build On: The Skeleton

Human *osteology*, from the Greek word for "bone" (*osteon*) and the suffix –*logy*, which means "to study," focuses on the 206 bones in the adult body's endoskeleton. But osteology is more than just bones; it's also ligaments and cartilage and the joints that make the whole assembly useful. In this chapter, you get lots of practice exploring the skeletal functions and how the joints work together.

Understanding the Functions of Dem Bones

REMEMBER

The skeletal system as a whole serves five key functions:

>> **Protection:** The skeleton encases and shields delicate internal organs that might otherwise be damaged during motion or crushed by the weight of the body itself. For example, the skull's cranium houses the brain, and the ribs and sternum of the thoracic cage protect the heart and lungs.

>> **Movement:** By providing anchor sites and a scaffold against which muscles can contract, the skeleton makes motion possible. The bones act as levers, the joints are the fulcrums, and the muscles apply the force. For instance, when the biceps muscle contracts, the radius and ulna bones of the forearm are lifted toward the humerus bone of the upper arm.

>> **Shape and Support:** The bones of the skeleton create the framework onto which the rest of our body is built, giving us our characteristic human shape. The density of the bones creates the rigidity needed to support all that tissue. The leg and foot bones are especially important for bearing the weight of our upright posture as is the curvature of our backbone.

>> **Mineral storage:** Calcium, phosphorous, and other minerals like magnesium must be maintained in the bloodstream at a constant level, so they're "banked" in the bones in case the dietary intake of those minerals drops. The bones' mineral content is constantly renewed, refreshing entirely about every nine months. A mere 35 percent decrease in blood calcium can cause convulsions.

>> **Blood cell formation:** Called *hematopoiesis,* or *hemopoiesis,* most blood cell formation in adults takes place within the red marrow inside the ends of long bones as well as within the pelvis, vertebrae, ribs, sternum, and cranial bones. Marrow produces three types of blood cells: *erythrocytes* (red cells), *leukocytes* (white cells), and *thrombocytes* (platelets). Most of these are formed in red bone marrow, although some types of white blood cells are produced in fat-rich yellow bone marrow. At birth, all bone marrow is red. With age, it converts to the yellow type. In cases of severe blood loss, the body can convert yellow marrow back to red marrow in order to increase blood cell production.

The following is an example of a question dealing with skeletal functions:

EXAMPLE

Q. Which of the following is not a function of the skeleton?

 a. Support of soft tissue

 b. Hemostasis

 c. Production of red blood cells

 d. Allow for movement

 e. Mineral storage

A. The correct answer is hemostasis, which is the stoppage of bleeding or blood flow. This is one of those frequent times when study of anatomy and physiology boils down to rote memorization of Latin and Greek roots (for help, check out the Cheat Sheet at www.dummies.com/cheatsheet/anatomyphysiologywb).

1 What is happening during hematopoiesis?

    a. Bone marrow is converting from red type to yellow type.

    b. Bone marrow is forming new blood cells.

    c. Bone marrow is removing damaged bone matrix.

    d. Bone marrow is converting back from yellow type to red type.

    e. Bone marrow is releasing minerals into the blood stream.

2 Which of the following IS NOT a role of minerals like phosphorous, calcium, and magnesium?

    a. To enhance the skeleton's structural integrity

    b. To provide cushioning and nourishment to the joints

    c. To ensure availability of these minerals

    d. To form networks within the bones' structure

3 In what way does the skeleton make locomotion possible?

    a. It ensures hematopoiesis occurs as needed.

    b. Muscles use the skeleton as a scaffold against which they can contract.

    c. It cushions the joints against erythrocytes.

    d. Bones generate the motion necessary for movement within the joints.

    e. It ensures the thoracic cavity remains properly filled.

Boning Up on Classifications and Structures

Adult bones are constituted of 35 percent protein (called *ossein*), 65 percent minerals (including calcium, phosphorus, and magnesium), and a small amount of water. Minerals give the bone strength and hardness. At birth, bones are soft and pliable because they are composed mostly of cartilage. As the body grows, older cartilage is gradually replaced by hard bone tissue. Bone density, mineral content, and collagen production decreases with age, causing the bones to become more brittle and easily fractured.

REMEMBER

Various types of bones make up the human skeleton, but fortunately for memorization purposes, bone type names match what the bones look like for the most part. They are as follows:

WARNING

>> **Long bones,** like those found in the arms and legs, defined by their length being greater than their width.

Though they aren't very long at all, the *phalanges* (finger bones) are, in fact, long bones

>> **Short bones,** such as those in the wrists (carpals) and ankles (tarsals), have a blocky structure and allow for a greater range of motion. *Sesamoid* bones include seed-shaped bones found in joints such as the *patella,* or kneecap.

>> **Flat bones,** such as the skull, sternum, and scapulae, shield soft tissues.

>> **Irregular bones,** such as the mandible (jawbone) and vertebrae, come in a variety of shapes and sizes suited for attachment to muscles, tendons, and ligaments.

Unfortunately for students of bone structures, there's no easy way to memorize them. So brace yourself for a rapid summary of what your textbook probably goes into in much greater detail.

Bones are comprised of two types of osseus tissue: *trabecular,* or *spongy,* bone and *compact* bone. These are found underneath the *periosteum,* which is a layer of dense irregular connective tissue that surrounds each bone. The periosteum assists in routing blood vessels deep into the bone as well as anchoring tendons and ligaments. It has both a vascular layer (the Latin word for "vessel" is *vasculum*) and an inner layer that contains the osteoblasts needed for bone growth and repair. A penetrating matrix of connective tissue, called *Sharpey's fibers,* connects the periosteum to the bone. Long bones also have an inner membrane, the *endosteum* that creates the *medullary cavity* that houses marrow. In flat, short, and irregular bones, spongy bone is found in the middle, surrounded by a layer of compact bone then wrapped in the periosteum. Long bones are a bit more complicated as they don't just grow in size but also change proportion.

The shaft of a long bone, the *diaphysis,* is mostly compact bone. It also contains the medullary cavity. The ends, or *epiphyses,* are encased in compact bone but contain mostly spongy bone. Between the two is the *metaphysis,* which contains the *epiphyseal plate.* This band, also referred to as the *growth plate,* is made of hyaline cartilage. This is the site of endochondral growth which allows long bones to increase in length. This process is discussed further in the next section, "Turning Bone into Bone."

Compact bone is highly organized with repeating units called *osteons*. As you read this section, refer to Figure 7-1 to help you put the description with the visual. An osteon is a series of concentric circles surrounding a circular, central canal called the *Haversian canal*. Each Haversian canal contains blood and lymphatic vessels as well as nerves. Outward from there are rings of matrix called *lamellae*. Bordering each lamella are several *lacunae*, little caves housing the bone cells, or *osteocytes*. Spreading through the lamellae are tiny tunnels called *canaliculi*, which connect the lacunae. Osteocytes are able to form gap junctions (we discuss these in Chapter 5) through the canaliculi, allowing resources to be passed cell-to-cell to those far away from the Haversian canal. The Haversian canals of each osteon are connected with *Volkmann's canals*, carrying the vessels deeper into the bone.

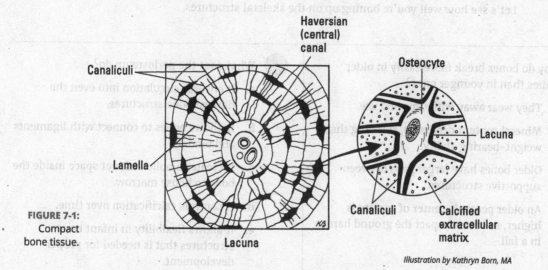

FIGURE 7-1:
Compact
bone tissue.

Illustration by Kathryn Born, MA

Spongy bone is much less organized. (Which means less things for you to label!) Rather than the matrix being built into rings, the calcium compounds are deposited onto collagen fibers. This forms the creation of bony bridge-like structures called *trabeculae*. While they do still have canaliculi (the little tunnels that connect cells to each other), there are no other canals or lacunae.

REMEMBER

In addition to all the internal structures of a bone, there are numerous external ones. These serve a variety of purposes, such as providing a point of attachment for tendons and ligaments and creating surfaces for *articulations* (joints; *articular* is an adjective describing areas related to movement between bones). The following describes the terms used to identify these features:

>> **Process:** A broad designation for any noticeable projection (example: mastoid process)

>> **Spine:** An abrupt or pointed projection (example: scapular spine)

>> **Trochanter:** A large, usually blunt process (example: greater trochanter of femur)

>> **Tubercle:** A smaller, rounded projection (example: lesser tubercle of humerus)

>> **Tuberosity:** A large, often rough process (example: tibial tuberosity)

>> **Crest:** A prominent ridge (example: iliac crest)

>> **Head:** A large, rounded of a bone, often set off from the shaft by a neck (example: head of femur)

>> **Condyle:** A rounded prominence that articulates with another bone (example: medial condyle of femur)

>> **Facet:** A small, smooth, and flat surface (example: facet of thoracic vertebrae)

>> **Fossa:** A deeper depression or pit (example: mandibular fossa)

>> **Sulcus:** A shallow groove (example: scapular notch)

>> **Foramen:** A hole or opening (example: foramen magnum of occipital bone)

>> **Meatus:** A canal or tunnel (example: external acoustic meatus of temporal bone)

Let's see how well you're boning up on the skeletal structures.

 4 Why do bones break more easily in older bodies than in younger ones?

a. They wear away with time and use.

b. Mineral content decreases, reducing their weight-bearing ability.

c. Older bones have larger gaps between supportive structures.

d. An older person's center of gravity is higher, so bones impact the ground harder in a fall.

 5 What does the periosteum do?

a. It ensures circulation into even the hardest bone structures.

b. It allows bones to connect with ligaments and tendons.

c. It creates a hollowed-out space inside the bone to house marrow.

d. It controls calcification over time.

e. It allows flexibility in infant bone structures that is needed for proper development.

 6 Where are Sharpey's fibers?

a. Inside Haversian canals

b. Woven through the spaces in trabecular bone

c. Between the periosteum and bone tissue

d. Behind any lingering mastoid tissue

e. Between cells in compact bone

 7 What is inside the medullary cavity?

a. Bone marrow

b. Epiphyses

c. Volkmann's canals

d. Trabecula

e. Osteoclasts

8 Volkmann's canals

a. are found in compact bone only. *(circled)*

b. contain the nutrient artery.

c. pass through the epiphysis.

d. form the center of an osteon.

e. supply blood to articulating cartilage.

9 Which type of bones forms the weight-bearing part of the skeleton?

a. Flat bones

b. Irregular bones

c. Long bones *(circled)*

d. Short bones

10-18 Classify the following bones by shape. Each classification may be used more than once.

a. Flat bone

b. Irregular bone

c. Long bone

d. Short bone

10 ___B___ Vertebrae of the vertebral column

11 ___C___ Femur in thigh

12 ___A___ Sternum

13 ___D___ Tarsals in ankle

14 ___C___ Humerus in upper arm

15 ___C___ Phalanges in fingers and toes

16 ___A___ Scapulae of shoulder

17 ___D___ Kneecap

18 ___B___ Carpals in wrist

19-31 Match the description with the bone landmarks or surface features.

a. Condyle

b. Crest

c. Facet

d. Foramen

e. Fossa

f. Head

g. Meatus

h. Process

i. Spine

j. Sulcus

k. Trochanter

l. Tubercle

m. Tuberosity

19 _____ An abrupt or pointed projection

20 _____ A large, usually blunt process

21 _____ A designation for any prominence or prolongation

22 _____ A large, often rough eminence

23 _____ A prominent ridge

24 _____ A large, rounded articular end of a bone; often set off from the shaft by the neck

25 _____ An oval articular prominence of a bone

26 _____ A smooth, flat, or nearly flat articulating surface

27 _____ A deeper depression

28 _____ A groove

29 _____ A hole

30 _____ A canal or opening to a canal

31 _____ A smaller, rounded eminence

 32-40 Use the terms that follow to identify the regions and structures of the long bone shown in Figure 7-2.

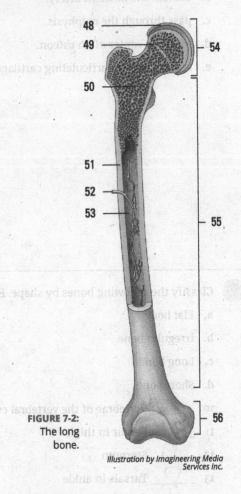

FIGURE 7-2:
The long bone.

Illustration by Imagineering Media Services Inc.

a. Diaphysis

b. Medullary cavity

c. Distal epiphysis

d. Spongy bone tissue

e. Medullary or nutrient artery

f. Proximal epiphysis

g. Epiphyseal plate

h. Articular cartilage

i. Compact bone tissue

Turning Bone into Bone: Ossification

There are three types of bone cells: *osteoblasts* (the builders), *osteoclasts* (the wreckers), and *osteocytes* (the maintenance workers). Osteocytes are mature bone cells, carrying out daily functions and influencing the other cells. The growth of bone tissue is carried out by osteoblasts through a process called *ossification*, of which there are two types: *endochondral* and *intramembranous*.

Endochondral ossification

During fetal development, most bones are first formed from hyaline cartilage through endochondral ossification. Over time, osteoblasts place calcium compounds (or other mineral salts) on the collagen fibers and remains of chondrocytes (cartilage cells). This *calcification* is the core of the ossification process. After birth, endochondral ossification begins in a secondary site — the epiphyseal plate. When *growth hormone* (GH) is released from the anterior pituitary, the chondrocytes in the plate are stimulated to divide and enlarge. This creates space within the bone that osteoblasts will calcify thus lengthening the bone. When osteoblasts are signaled to stop building (or because they've walled themselves off in a lacuna), they become osteocytes and they remain as such for the remainder of their lives.

Intramembranous ossification

Intramembranous ossification occurs during fetal development in many flat bones. They begin as a fibrous membrane and osteoblasts calcify the tissue from the center of the bone outward. As such, the edges of the skull bones are not yet calcified at birth; this allows for the baby's skull to compress, in a sense, during childbirth (which is also why many babies come out looking like cone heads). The soft spot between the bones is called a *fontanel* and will continue ossification over the next year.

As development continues, flat, short, and irregular bones utilize this method to grow in size. Intramembranous growth occurring later in development like this is sometimes referred to as *appositional growth*. A thin layer of connective tissue (which can be cartilage but doesn't have to be) is built just beneath the periosteum and is then calcified by osteoblasts. The bone grows proportionally one layer, or membrane, at a time. In long bones, growth in width occurs in this manner, only the tissue is built outside of the endosteum instead.

In either process, osteoblasts will cease calcification once spongy bone has been formed. To build compact bone, new osteoblasts will be brought in to fill in the empty space starting near a blood vessel, building out one ring at a time (this is how a lamella is formed). Once your growth plate is closed, or fully calcified, you are unable to gain any height but your bones are never really done developing. They are constantly undergoing a process called *remodeling*. As we apply forces through exercise, or just our daily activities, the bone matrix weakens (picture a brand new brick wall compared to an old, weathered one). Osteoclasts are signaled to begin the remodeling process. They will secrete acid to dissolve the area's matrix. The calcium compounds are broken up and the calcium ions return to the bloodstream. After the osteoclasts clean up the area via phagocytosis, osteoblasts will move in and deposit new calcium compounds, thereby strengthening the bone.

REMEMBER Remodeling does not occur one bone at a time (like road construction) or even in large chunks. It is done everywhere, in little bits, all the time. This way, the structural integrity of the bone is never put in jeopardy.

Remodeling is also under hormonal control. Since the bones serve as a reservoir for minerals, calcium in particular, the body can trigger osteoblasts or osteoclasts depending on the need. When calcium in the blood is in abundance, the thyroid gland releases *calcitonin*, triggering the osteoblasts to create calcium compounds and deposit them into the matrix. When blood calcium levels are low, the parathyroid gland releases *parathyroid hormone* (PTH) to signal the osteoclasts to break down matrix. Calcium ions are released into the bloodstream so normal functioning can continue; calcium ions are particularly important for muscle contraction as well as nervous communication.

41 What does calcitonin do?

a. It regulates metabolism of calcium.

b. It controls the development of Haversian systems.

c. It influences the formation of bone marrow.

d. It breaks down calcium compounds in the matrix.

e. It encourages the proper layering of osteoclasts.

42-51 Fill in the blanks to complete the following sentences:

During fetal development, most bones are first laid down as 42._____ during the fifth week after conception.

Development of the bone begins with 43._____, the depositing of mineral compounds. This is performed by 44._____ that attach themselves to the cartilage.

Ossification in long bones begins in the

45._____ and moves toward the 46._____ of the bone. These areas remain separated by a layer of uncalcified cartilage called the 47._____.

Another bone cell that enters with the blood supply is the 48._____, which helps absorb the remnants of original cartilage as ossification occurs. Later it helps absorb bone tissue from the center of the long bone's shaft, forming the 49._____ cavity. Unlike bones in the rest of the body, those of the skull and mandible (lower jaw) are first laid down as 50._____. In the skull, the edges of the bone don't ossify in the fetus but remain membranous and form 51._____.

52 The constant breakdown and rebuilding of bone matrix is termed

a. endochondral growth.

b. primary calcification.

c. ossification.

d. intramembranous growth.

e. remodeling.

53 Endochondral growth is triggered by which hormone?

a. PTH

b. Calcitonin

c. Calcitriol

d. GH

e. Osteocalcin

54 Identify the differences between intramembranous and endochondral growth.

_____

_____

_____

_____

_____

The Axial Skeleton: Keeping It All in Line

Just as the Earth rotates around its axis, the axial skeleton lies along the midline, or center, of the body. Think of your spinal column and the bones that connect directly to it — the rib (thoracic) cage and the skull. The tiny *hyoid* bone, which lies just above your larynx, or voice box, also is considered part of the axial skeleton, although it's the only bone in the entire body that doesn't articulate, or connect, with any other bone. It's also known as the *tongue bone* because the tongue's muscles attach to it. There are a total of 80 named bones in the axial skeleton, which supports the head and trunk of the body and serves as an anchor for the pelvic girdle.

Making a hard head harder

Of the 80 named bones in the axial skeleton, 29 are in (or very near) the skull. In addition to the hyoid bone, 8 bones form the cranium to house and protect the brain, 14 form the face, and 6 bones make it possible for you to hear.

Fortunately for the cramming student, most of the bones in the skull come in pairs. In the cranium there's just one of each of the following:

>> *Frontal* bone (forehead)

>> *Occipital* bone (back and base of the skull) containing occipital condyles, which articulate with the atlas of the vertebral column

>> *Ethmoid* bone (made of several plates, or sections, between the eye orbits in the nasal cavity)

>> *Sphenoid* bone (a butterfly-shaped structure that forms the floor of the cranial cavity and the backs of the orbits)

But there are two *temporal* bones each housing: (1) the hearing organs in the *auditory meatus*; (2) a styloid process, a long pointed process for anchoring muscles; (3) a mandibular fossa articulating with the condyle of mandible; and (4) a zygomatic process that projects anteriorly joining the zygomatic bone forming the cheek prominence. There are two *parietal* bones (roof and sides of the skull). These bones are attached along *sutures* called the following:

>> *Coronal* (located at the top of the skull between the two parietal bones)

>> *Squamosal* (located on the sides of the head surrounding the temporal bone)

>> *Sagittal* (along the midline atop the skull located between the two parietal bones)

>> *Lambdoidal* (forming an upside-down V — the shape of the Greek letter lambda — on the back of the skull)

In the face, there is only one *mandible* (jawbone) and one *vomer* dividing the nostrils, but there are two each of *maxillary* (upper jaw), *zygomatic* (cheekbone), *nasal*, *lacrimal* (a small bone in the eye socket), *palatine* (which makes up part of the eye socket, nasal cavity, and roof of the mouth), and *inferior nasal concha*, or turbinated, bones. The mandible has two lateral condyles that articulate in the mandibular fossa of the temporal bone. (The maxillary bones are also called, simply, the *maxilla*.) Each ear contains three ossicles, or bonelets, which also happen to be the smallest bones in the human body: the *malleus*, *incus*, and *stapes*.

The floor of the cranial cavity contains several openings, or *foramina* (the singular is *foramen*), that allow various nerves and vessels to connect to the brain.

>> The holes in the ethmoid bone's *cribriform plate* are *olfactory foramina* that allow olfactory — or sense of smell — receptors to pass through to the brain. A process called the *crista galli* extends into the brain cavity for the attachment of the *meninges* (outer membranes) of the brain.

>> A large hole in the occipital bone called the *foramen magnum* allows the spinal cord to connect with the brain.

>> The sphenoid bone is riddled with foramina.

• The *optic foramen* allows passage of the optic nerves.

• The *jugular foramen* allows passage of the jugular vein and several cranial nerves.

- The *foramen rotundum* allows passage of the *trigeminal nerve,* which is the chief sensory nerve to the face and controls the motor functions of chewing.

- The *foramen ovale* allows passage of the nerves controlling the tongue, among other things.

- The *foramen spinosum* allows passage of the middle *meningeal* artery, which supplies blood to various parts of the brain.

- The sphenoid bone also features the *sella turcica,* or Turk's saddle, that cradles the pituitary gland and forms part of the *foramen lacerum,* which allows passage of several key components of the autonomic nervous system.

Encased within the frontal, sphenoid, ethmoid, and maxillary bones of the skull are several air-filled, mucous-lined cavities called *paranasal sinuses* (named for the bones in which they are contained). Although you may think their primary function is to drive you crazy with pressure and infections, the sinuses actually lighten the skull's weight. They make it easier to hold your head up high; they warm and humidify inhaled air; and they act as resonance chambers to prolong and intensify the reverberations of your voice. These drain into the nose and cause so much trouble when you cry or have a cold. The temporal bone contains the *mastoid sinuses,* which drain into the middle ear (hence the earache referred to as *mastoiditis*).

Putting your backbones into it

REMEMBER

The axial skeleton also consists of 33 bones in the vertebral column, laid out in four distinct curvatures, or areas.

>> **The cervical, or neck, curvature** has seven vertebrae, with the *atlas* and *axis* bones positioned in the first and second spots, respectively, forming the joint connecting the skull and vertebral column that allows you to turn your head. The axis, or second cervical vertebra, contains the *dens,* or *odontoid process.* The atlas rotates around the dens, turning the head. (In a sense, the atlas bone holds the world of the head on its shoulders, as the Greek god Atlas held the Earth.)

>> **The thoracic, or chest, curvature** has 12 vertebrae that articulate with the ribs, most of which attach to the sternum anteriorly by costal cartilage, forming the rib cage that protects the heart and lungs.

>> **The lumbar, or small of the back, curvature** contains five vertebrae and carries most of the weight of the body, which means that it generally suffers the most stress.

>> **The sacral, or pelvic, curvature** includes the five fused vertebrae of the *sacrum* anchoring the pelvic girdle by the *sacroiliac joint* and four fused vertebrae of the *coccyx,* or tailbone.

REMEMBER

The spinal cord extends down the center of the vertebrae only from the base of the brain to the uppermost lumbar vertebrae.

Each vertebra, except the atlas, consists of a body and a vertebral arch, which features a long dorsal projection called a *spinous process* that provides a point of attachment for muscles and ligaments. On either side of this are the *laminae,* broad plates of bone on the posterior surface that form a bony covering over the spinal canal. The laminae attach to the two *transverse processes,* which in turn are attached to the body of the vertebra by regions called the *pedicles.* The vertebrae align to form a large opening, called the *vertebral foramen,* allowing the passage of the spinal cord. Laterally, between the vertebrae, are openings called the *intervertebral foramina* that allow the spinal nerves to exit the vertebral column. The vertebra have superior articulating facets that articulate with inferior articulating facets of the adjacent vertebra, increasing the rigidity of the column, making the backbone more stable. Fibrocartilage discs located between the vertebrae act as shock absorbers. Openings in the transverse process, called the *transverse foramina*, allow large vessels and nerves ascending the neck to reach the brain.

Connecting to the vertebral column are the 12 pairs of ribs that make up the thoracic cage. All 12 pairs attach to the thoracic vertebrae, but the first 7 pairs attach to the *sternum*, or breastbone, by *costal cartilage;* they're called *true ribs.* Pairs 8, 9, and 10 attach to the cartilage of the seventh pair, which is why they're called *false ribs.* The last two pairs aren't attached in front at all, so they're called *floating ribs.*

The sternum has three parts:

>> **Manubrium:** The superior region that articulates with the clavicle, at the clavicular notch, and the first two pairs of ribs located up top, where you can feel a jugular notch in your chest in line with your *clavicles,* or collarbones.

>> **Body:** The middle part of the sternum forms the bulk of the breastbone and has notches on the sides where it articulates with the third through seventh pairs of ribs.

>> **Xiphoid process:** The lowest part of the sternum is an attachment point for the diaphragm and some abdominal muscles. (Interesting fact: Emergency medical technicians learn to administer CPR at least three finger widths above the xiphoid.)

 Use the terms that follow to identify the bones, sutures, and landmarks of the skull shown in Figure 7-3.

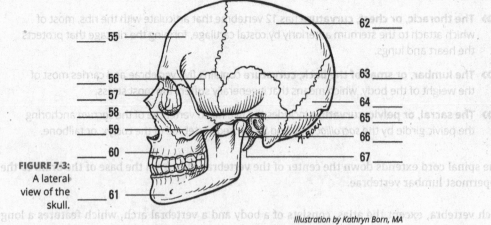

FIGURE 7-3:
A lateral
view of the
skull.

_____ 55

_____ 56

_____ 57

_____ 58

_____ 59

_____ 60

_____ 61

62 _____

63 _____

64 _____

65 _____

66 _____

67 _____

Illustration by Kathryn Born, MA

a. Temporal bone

b. Sphenoid bone

c. Nasal bone

d. Styloid process

e. Frontal bone

f. Ethmoid bone

g. Maxilla

h. Parietal bone

i. Occipital bone

j. Zygomatic bone

k. Mandibular condyle

l. Zygomatic process

m. Mandible

Match the skull bones with their connecting sutures.

a. Squamosal suture

b. Lambdoidal suture

c. Sagittal suture

d. Coronal suture

68 _____ Frontal and parietals

69 _____ Occipital and parietals

70 _____ Parietal and parietal

71 _____ Temporal and parietal

72-78 Use the terms that follow to identify the bones and landmarks of the skull shown in Figure 7-4.

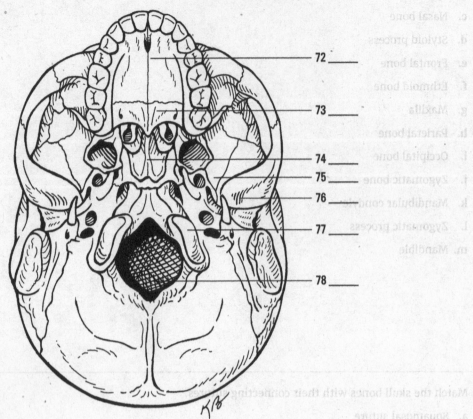

72 _____

73 _____

74 _____

75 _____

76 _____

77 _____

78 _____

FIGURE 7-4:
Inferior view
of the skull.

Illustration by Kathryn Born, MA

a. Vomer

b. Mandibular fossa

c. Foramen magnum

d. Palatine bone

e. Occipital condyle

f. Maxilla

g. Sphenoid bone

79-82 Use the terms that follow to identify the sinuses shown in Figure 7-5.

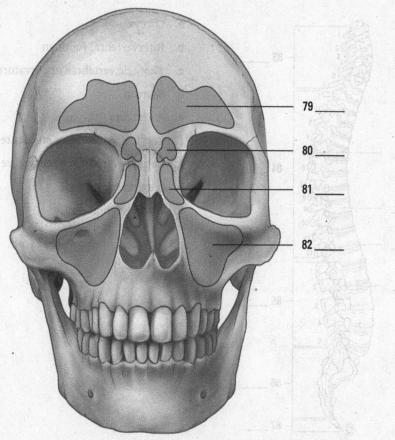

79 _____

80 _____

81 _____

82 _____

FIGURE 7-5:
Sinus view
of the skull.

Illustration by Imagineering Media Services Inc.

    a. Sphenoid sinus

    b. Frontal sinus

    c. Maxillary sinus

    d. Ethmoid sinus

83-91 Use the terms that follow to identify the regions, structures, and landmarks of the vertebral column shown in Figure 7-6.

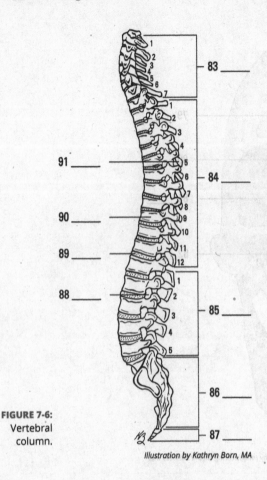

a. Coccyx

b. Intervertebral foramen

c. Thoracic vertebrae or curvature

d. Sacrum

e. A vertebra

f. Cervical vertebrae or curvature

g. Lumbar vertebrae or curvature

h. A spinous process

i. Intervertebral disc

FIGURE 7-6: Vertebral column.

Illustration by Kathryn Born, MA

92-97 Fill in the blanks to complete the following sentences:

The organs protected by the thoracic cage include the 92. _____ and the

93. _____. The first seven pairs of ribs attach to the sternum by the costal cartilage

and are called 94. _____ ribs. Pairs 8 through 10 attach to the costal cartilage of

the seventh pair and not directly to the sternum, so they're called 95. _____ ribs.

The last two pairs, 11 and 12, are unattached anteriorly, so they're called 96. _____

ribs. There's one bone in the entire skeleton that doesn't articulate with any other bones but nonetheless

is considered part of the axial skeleton. It's called the 97. _____ bone.

_____

The Appendicular Skeleton: Reaching beyond Our Girdles

Whereas the axial skeleton (described earlier in this chapter) lies along the body's central axis, the appendicular skeleton's 126 bones include those in all four appendages — arms and legs — plus the two primary girdles to which the appendages attach: the pectoral (chest) girdle and the pelvic (hip) girdle.

The pectoral girdle is made up of a pair of *clavicles*, or collarbones, which attach to the sternum medially and to the scapula laterally articulating with the *acromion process*, a bony prominence at the top of each of the pair of scapulae, better known as shoulder blades. Each scapula has a depression in it called the *glenoid fossa*, where the head of the *humerus* (upper arm bone) is attached. At the opposite end, the *capitulum* of the humerus articulates with the *trochlea* of the forearm's long *ulna* bone to form the elbow joint. The process called the *olecranon* forms the elbow and is also referred to as the funny bone, although banging it into something usually feels anything but funny. The forearm also contains a bone called the *radius*. The biceps muscle attaches to the radial tuberosity, flexing the elbow. The radius, together with the ulna, articulates with the eight small *carpal* bones that form the wrist. The carpals articulate with the five *metacarpals* that form the hand, which in turn connect with the *phalanges* (finger bones), which are found as a pair in the thumb and as triplets in each of the fingers.

The pelvic girdle consists of two hipbones, called *os coxae*, as well as the *sacrum* and *coccyx*, more commonly referred to as the tailbone. During early developmental years, the os coxa consists of three separate bones — the *ilium*, the *ischium*, and the *pubis* — that later fuse into one bone sometime between the ages of 16 and 20. Posteriorly, the os coxa articulates with the sacrum, forming the *sacroiliac joint*, the source of much lower back pain; it's formed by the connection of the hipbones at the sacrum. Toward the front of the pelvic girdle, the two os coxae join to form the *symphysis pubis*, which is made up of fibrocartilage. A cuplike socket called the *acetabulum* articulates with the ball-shaped head of the *femur* (thigh bone). The femur is the longest bone in the body. The femur articulates with the *tibia* (shin bone) at the knee, which is covered by the *patella* (kneecap). Also inside each lower leg is the *fibula* bone, which joins with the tibia to connect with the seven *tarsal* bones that make up the ankle. These are the calcaneus (heel bone), talus, navicular, cuboid, lateral cuneiform, intermediate cuneiform, and medial cuneiform. The largest of these bones is the calcaneus. The tarsals join with the five *metatarsals* that form the foot, which in turn connect to the *phalanges* of the toes — a pair of phalanges in the big toe and triplets in each of the other toes.

 Use the terms that follow to identify the bones and structures of the appendicular skeleton shown in Figure 7-7.

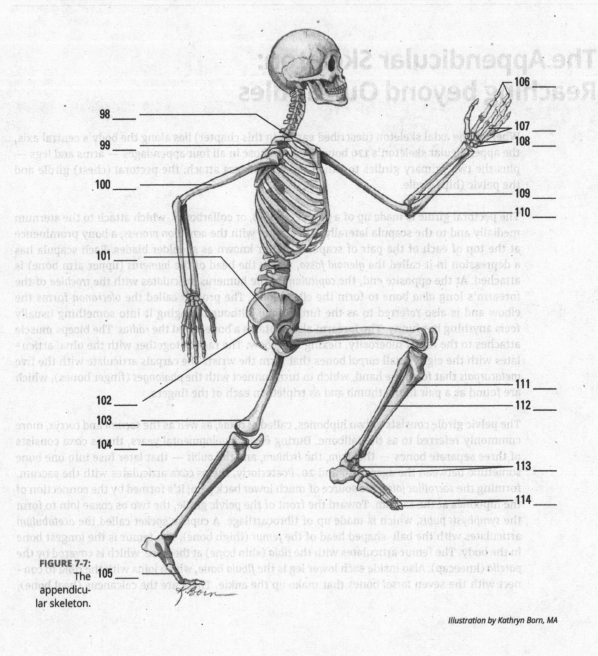

FIGURE 7-7: The appendicular skeleton.

Illustration by Kathryn Born, MA

a. Tibia

b. Ulna

c. Scapula

d. Metatarsals

e. Carpals

f. Phalanges of the foot

g. Clavicle

h. Fibula

i. Patella

j. Humerus

k. Radius

l. Os coxa

m. Sacrum

n. Phalanges of the hand

o. Metacarpals

p. Tarsals

q. Femur

---

 115 Where do the clavicles articulate with the scapulae?

a. At the acromion process

b. Below the glenoid fossa

c. Behind the coracoid process

d. At the acetabulum

e. Along the upper spine

116 Where does the biceps muscle attach on the radius?

a. The radial notch

b. The humeral capitulum

c. The ulnar trochlea

d. The radial tuberosity

e. The ulnar notch

117 Which of these bones IS NOT part of the pelvic girdle?

   a. Ilium

   b. Lumbar vertebrae

   c. Sacrum

   d. Ischium

118 What is the formal term for the prominence commonly referred to as the elbow?

   a. Olecranon process

   b. Trochlear notch

   c. Capitulum

   d. Radial notch

   e. Coronoid process

119 What is the formal term for the socket at the head of the femur?

   a. Obturator foramen

   b. Acetabulum

   c. Ischial tuberosity

   d. Symphysis pubis

   e. Greater sciatic notch

120 Which is largest and strongest tarsal bone?

   a. The talus

   b. The cuboid

   c. The cuneiform

   d. The calcaneus

   e. The metatarsal

121 Where are people who complain about their sacroiliac experiencing pain?

    a. Lower back

    b. Neck

    c. Feet

    d. Knees

    e. Hands

Arthrology: Articulating the Joints

Arthrology, which stems from the ancient Greek word *arthros* (meaning "jointed"), is the study of those structures that hold bones together, allowing them to move to varying degrees — or fixing them in place — depending on the design and function of the joint. The term *articulation*, or *joint*, applies to any union of bones, whether it moves freely or not at all. As such, joints can be classified either by structure or movement.

REMEMBER

The movement classification of joints are as follows:

  » **Synarthrotic** — immovable

  » **Amphiarthrotic** — slightly movable (more like have "give" or allow compression)

  » **Diarthrotic** — highly movable

The three structural types of joints are as follows:

  » **Fibrous:** These are characterized by the fibrous connective tissue that joins the two bones. Most of these are synarthrotic and come in three forms:

    • *Sutures* occur when the bones are joined by a piece of dense connective tissue. These can be found between skull bones.

    • *Gomphosis* is found where the teeth attach to the sockets of the mandible and maxilla. They are joined by the *periodontal ligament* (although teeth aren't actually bones).

    • *Syndesmosis* is the only amphiarthrotic fibrous joint. Bones are joined by a sheet of connective tissue, like that of the articulations between the tibia and fibula. It allows for stretch and even some twisting.

  » **Cartilaginous:** This type of joint is found in two forms:

    • *Synchondrosis* is formed from rigid, hyaline cartilage that allows no movement. The most common example is the epiphyseal plate of the long bone. Other examples are the joint between the ribs, costal cartilage, and sternum.

    • *Symphysis* occurs where fibrocartilage pads connect bones. Since the pads are flexible, these are amphiarthrotic (think about how you can arch your back). Examples include the intervertebral discs and the symphysis pubis.

» **Synovial:** All synovial joints are diarthrotic but to varying degrees. They are surrounded by a *joint capsule* that isolates the articulation leaving space that is filled with *synovial fluid* to reduce friction. Each adjoining bone is capped with hyaline cartilage called the *articular cartilage* and stabilized with ligaments. Some joints such as the knee and wrist have an additional pad of fibrocartilage called a *meniscus* between the bones for shock absorption. Many synovial joints also contain *bursae,* which are fluid filled sacs to reduce friction as tendons rub past the bones during movement.

There are six classifications of synovial joints:

» **Gliding (or plane):** Curved or flat surfaces slide against one another, such as between the carpal bones in the wrist or between the tarsal bones in the ankle. Allows movement in two planes and rotation is possible but limited by their small size and ligament connections.

» **Hinge:** A convex surface joints with a concave surface, allowing right-angle motions in one plane, such as elbows, knees, and joints between the phalanges.

» **Pivot:** One bone pivots or rotates around a stationary bone, such as the atlas rotating around the dens of the axis at the top of the vertebral column and the radioulnar joint. Rotation is the only type of movement allowed.

» **Condyloid:** The oval head of one bone fits into a shallow depression in another, as in the carpal-metacarpal joint at the wrist and the tarsal-metatarsal joint at the ankle. Movement is allowed through all three planes but no rotation.

» **Saddle:** Each of the adjoining bones is shaped like a saddle (the technical term is *reciprocally concavo-convex*), like in the carpometacarpal joint of the thumb (where the thumb meets the wrist). Movement is allowed through two planes and no rotation.

» **Ball-and-socket:** The round head of one bone fits into a cuplike cavity in the other bone, allowing movement in all planes of motion plus rotation. These include your two most movable joints — the shoulder between the humerus and scapula and the hip between the femur and the os coxa.

REMEMBER

The following are the main types of joint movement:

» **Flexion:** A decrease in the angle between two bones

» **Extension:** An increase in the angle between two bones

» **Abduction:** Movement away from the midline of the body

» **Adduction:** Movement toward the midline of the body

» **Rotation:** Turning around an axis (medial or lateral)

» **Pronation:** Downward or palm downward

» **Supination:** Upward or palm upward

» **Eversion:** Turning of the sole of the foot outward

» **Inversion:** Turning of the sole of the foot inward

» **Circumduction:** Forming a cone with the arm or leg (think tracing a circle)

Match the articulations with their joint types. Some joint types may be used more than once.

a. Fibrous joint

b. Cartilaginous joint-synchondrosis

c. Cartilaginous joint-symphysis

d. Synovial joint

122 _____ Sutures of the skull

123 _____ Between carpals

124 _____ Knee joint

125 _____ Symphysis pubis

126 _____ Epiphyseal plate

127 _____ Intervertebral discs

 Use the terms that follow to identify the structures that form the synovial joint shown in Figure 7-8.

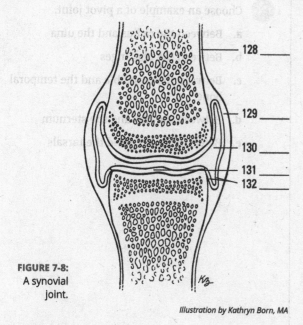

FIGURE 7-8:
A synovial
joint.

Illustration by Kathryn Born, MA

a. Synovial (joint) cavity

b. Periosteum

c. Synovial membrane

d. Articular cartilage

e. Joint (fibrous) capsule

133 What is the formal term for an immovable joint?

a. An amphiarthrosis

b. A synarthrosis

c. A syndesmosis

d. A diarthrosis

e. A synchondrosis

134 What material or structure allows free movement in a joint?

a. The bursa

b. The periosteum

c. Synovial fluid

d. The meniscus

e. Bone marrow

 135 Choose an example of a ball-and-socket joint.

a. Between the finger and hand

b. The hip

c. The ankle

d. The elbow

e. Between the wrist bones

 136 Choose an example of a pivot joint.

a. Between the radius and the ulna

b. Between the phalanges

c. Between the mandible and the temporal bone

d. Between the ribs and the sternum

e. Between the tibia and the tarsals

137 What is the structure in the knee that divides the synovial joint into two separate compartments?

    a. The bursa

    b. Joint fat

    c. The tendon sheath

    d. The collateral ligaments

    e. The meniscus, or articular disc

138–147 Match the type of joint movement with its description.

    a. Upward or palm upward

    b. Decrease in the angle between two bones

    c. Turning of the sole of the foot inward

    d. Downward or palm downward

    e. Increase in the angle between two bones

    f. Turning of the sole of the foot outward

    g. Movement away from the midline of the body

    h. Forming a cone with the arm or leg

    i. Turning around an axis

    j. Movement toward the midline of the body

    138 _____ Flexion

    139 _____ Extension

    140 _____ Abduction

    141 _____ Adduction

    142 _____ Rotation

    143 _____ Pronation

    144 _____ Supination

    145 _____ Eversion

    146 _____ Inversion

    147 _____ Circumduction

Answers to Questions on the Skeleton

The following are answers to the practice questions presented in this chapter.

1. What is happening during hematopoiesis? **b. Bone marrow is forming new blood cells.** Most of the other answers are nonsense.

2. Which of the following IS NOT a role of minerals like phosphorous, calcium, and magnesium? **b. To provide cushioning and nourishment to the joints.** The minerals in the bone provide strength and bind together to organize the structure of the matrix, thus Choices a and d are both accurate. Bones also store the minerals — if too little is consumed, hormones can trigger the release of them from the bones.

3. In what way does the skeleton make locomotion possible? **b. Muscles use the skeleton as a scaffold against which they can contract.** You'd be a motionless blob without a skeleton to support coordinated movement, and, besides, the rest of the answers are utter nonsense.

4. Why do bones break more easily in older bodies than in younger ones? **b. Mineral content decreases, reducing their weight-bearing ability.** The correct answer becomes obvious through a process of elimination.

5. What does the periosteum do? **b. It allows bones to connect with ligaments and tendons.** Back to Greek again: *peri* means "around" and *osteon* means "bone," so the periosteum is "around the bone."

6. Where are Sharpey's fibers? **c. Between the periosteum and bone tissue.** Described by anatomist William Sharpey in 1846, these are also called *perforating fibers*. They connect the periosteum to the matrix of the bone tissue underneath.

7. What is inside the medullary cavity? **a. Bone marrow.** Usually you'll find yellow marrow, although in infants red marrow also is present.

8. Volkmann's canals **a. are found in compact bone only.** Ironically, anatomist Alfred Wilhelm Volkmann was most noted for his observations of the physiology of the nervous system, not bones.

9. Which type of bones forms the weight-bearing part of the skeleton? **c. Long bones.** After all, they are the largest and most robust of the body's bones.

10. Vertebrae of the vertebral column: **b. Irregular bone**

11. Femur in thigh: **c. Long bone**

12. Sternum: **a. Flat bone**

13. Tarsals in ankle: **d. Short bone**

14. Humerus in upper arm: **c. Long bone**

15. Phalanges in fingers and toes: **c. Long bone**

16. Scapulae of shoulder: **a. Flat bone**

17. Kneecap: **d. Short bone** (or to be more specific, sesamoid)

18. Carpals in wrist: **d. Short bone**

(19) An abrupt or pointed projection: **i. Spine**

(20) A large, usually blunt process: **k. Trochanter**

(21) A designation for any prominence or prolongation: **h. Process**

(22) A large, often rough eminence: **m. Tuberosity**

(23) A prominent ridge: **b. Crest**

(24) A large, rounded articular end of a bone, often set off from the shaft by the neck: **f. Head**

(25) An oval articular prominence of a bone: **a. Condyle**

(26) A smooth, flat, or nearly flat articulating surface: **c. Facet**

(27) A deeper depression: **e. Fossa**

(28) A groove: **j. Sulcus**

(29) A hole: **d. Foramen**

(30) A canal or opening to a canal: **g. Meatus**

(31) A smaller, rounded eminence: **l. Tubercle**

(32-40) Following is how Figure 7-2, the long bone, should be labeled.

> 32. h. Articular cartilage; 33. g. Epiphyseal plate; 34. d. Spongy bone; 35. i. Compact bone tissue; 36. e. Medullary or nutrient artery; 37. b. Medullary cavity; 38. f. Proximal epiphysis; 39. a. Diaphysis; 40. c. Distal epiphysis

(41) What does calcitonin do? **a. It regulates metabolism of calcium.** Once you remember that calcitonin comes from the metabolism's pacemaker, the thyroid gland, you can discern the right answer.

(42-51) Fill in the blanks to complete the following sentences:

> Bones are first laid down as **42. cartilage** during the fifth week after conception. Development of the bone begins with **43. calcification**, the depositing of calcium and phosphorus. Next, the blood supply entering the cartilage brings **44. osteoblasts** that attach themselves to the cartilage. Ossification in long bones begins in the **45. diaphysis** of the long bone and moves toward the **46. epiphysis** of the bone. The epiphyseal and diaphyseal areas remain separated by a layer of uncalcified cartilage called the **47. epiphyseal plate.**

> Another very large cell that enters with the blood supply is the **48. osteoclast,** which helps absorb the cartilage as ossification occurs. Later it helps absorb bone tissue from the center of the long bone's shaft, forming the **49. medullary or marrow** cavity. Unlike bones in the rest of the body, those of the skull and mandible (lower jaw) are first laid down as **50. membrane.** In the skull, the edges of the bone don't ossify in the fetus but remain membranous and form **51. fontanels.**

(52) The constant breakdown and rebuilding of bone matrix is termed **e. remodeling.** Just think about remodeling a room of a house, you have to get rid of the stuff you don't want first.

(53) Endochondral growth is triggered by which hormone? **d. GH.** Growth hormone (GH) triggers bone growth — easy to remember right? Calcitonin and calcitriol specifically trigger calcification by osteoblasts, and PTH (parathyroid hormone) is the only hormone that triggers osteoclasts. Osteocalcin is a structural protein, not a hormone.

54 Identify the differences between intramembranous and endochondral growth. **These two processes share more in common than differences. Once osteoblasts are triggered to deposit mineral compounds, the steps are the same. The difference lies in how they create the space. Endochondral growth uses hyaline cartilage as a template while intramembranous uses fibrous connective tissue. In fetal development, all bones are formed via endochondral ossification with the exception of the flat bones of the skull and the mandible (which utilize endochondral). Throughout life, only long bones use endochondral growth (to increase their length). All other bones use intramembranous growth to increase in size proportionally, one layer at a time. Long bones do this to increase their width.**

55-67 Following is how Figure 7-3, the lateral view of the skull, should be labeled.

55. e. **Frontal bone**; 56. b. **Sphenoid bone**; 57. c. **Nasal bone**; 58. f. **Ethmoid bone**; 59. j. **Zygomatic bone**; 60. g. **Maxilla**; 61. m. **Mandible**; 62. h. **Parietal bone**; 63. a. **Temporal bone**; 64. i. **Occipital bone**; 65. l. **Zygomatic process**; 66. k. **Mandibular condyle**; 67. d. **Styloid process**

68 Frontal and parietals: d. **Coronal suture**

69 Occipital and parietals: b. **Lambdoidal suture**

70 Parietal and parietal: c. **Sagittal suture**

71 Temporal and parietal: a. **Squamosal suture**

72-78 Following is how Figure 7-4, the inferior view of the skull, should be labeled.

72. f. **Maxilla**; 73. d. **Palatine bone**; 74. a. **Vomer**; 75. g. **Sphenoid bone**; 76. b. **Mandibular fossa**; 77. e. **Occipital condyle**; 78. c. **Foramen magnum**

79-82 Following is how Figure 7-5, the sinus view of the skull, should be labeled.

79. b. **Frontal sinus**; 80. d. **Ethmoid sinus**; 81. a. **Sphenoid sinus**; 82. c. **Maxillary sinus**

83-91 Following is how Figure 7-6, the vertebral column, should be labeled.

83. f. **Cervical vertebrae or curvature**; 84. c. **Thoracic vertebrae or curvature**; 85. g. **Lumbar vertebrae or curvature**; 86. d. **Sacrum**; 87. a. **Coccyx**; 88. b. **Intervertebral foramen**; 89. i. **Intervertebral disc**; 90. e. **A vertebra**; 91. h. **A spinous process**

92-97 Fill in the blanks to complete the following sentences:

The organs protected by the thoracic cage include the 92. **heart** and the 93. **lungs**. The first seven pairs of ribs attach to the sternum by the costal cartilage and are called 94. **true** ribs. Pairs 8 through 10 attach to the costal cartilage of the seventh pair and not directly to the sternum, so they're called 95. **false** ribs. The last two pairs, 11 and 12, are unattached anteriorly, so they're called 96. **floating** ribs. There's one bone in the entire skeleton that doesn't articulate with any other bones but nonetheless is considered part of the axial skeleton. It's called the 97. **hyoid** bone.

98-114 Following is how Figure 7-7, the frontal view of the skeleton, should be labeled.

98. g. **Clavicle**; 99. c. **Scapula**; 100. j. **Humerus**; 101. l. **Os coxa**; 102. m. **Sacrum**; 103. q. **Femur**; 104. i. **Patella**; 105. d. **Metatarsals**; 106. n. **Phalanges of the hand**; 107. o. **Metacarpals**; 108. e. **Carpals**; 109. k. **Radius**; 110. b. **Ulna**; 111. a. **Tibia**; 112. h. **Fibula**; 113. p. **Tarsals**; 114. f. **Phalanges of the foot**

115 Where do the clavicles articulate with the scapulae? a. **At the acromion process**

116. Where does the biceps muscle attach on the radius? **d. The radial tuberosity**

117. Which of these bones is not part of the pelvic girdle? **b. Lumbar vertebrae**

118. What is the formal term for the prominence commonly referred to as the elbow? **a. Olecranon process**

119. What is the formal term for the socket at the head of the femur? **b. Acetabulum**

120. Which is the largest and strongest tarsal bone? **d. The calcaneus**

121. Where are people who complain about their sacroiliac experiencing pain? **a. Lower back**

122. Sutures of the skull: **a. Fibrous joint**

123. Between carpals: **d. Synovial joint**

124. Knee joint: **d. Synovial joint**

125. Symphysis pubis: **c. Cartilaginous joint-symphysis**

126. Epiphyseal plate: **b. Cartilaginous joint-synchondrosis**

127. Intervertebral discs: **c. Cartilaginous joint-symphysis**

128–132. Following is how Figure 7-8, the synovial joint, should be labeled.

128. **b. Periosteum**; 129. **e. Fibrous capsule**; 130. **a. Synovial (joint) cavity**; 131. **d. Articular cartilage**; 132. **c. Synovial membrane**

133. What is the formal term for an immovable joint? **b. A synarthrosis**

134. What material or structure allows free movement in a joint? **c. Synovial fluid**

135. Choose an example of a ball-and-socket joint. **b. The hip**

136. Choose an example of a pivotal joint. **a. Between the radius and the ulna**

137. What is the term for the structure in the knee that divides the synovial joint into two separate compartments? **d. Meniscus or articular disc**

138. Flexion: **b. Decrease in the angle between two bones**

139. Extension: **e. Increase in the angle between two bones**

140. Abduction: **g. Movement away from the midline of the body**

141. Adduction: **j. Movement toward the midline of the body**

142. Rotation: **i. Turning around an axis**

143. Pronation: **d. Downward or palm downward**

144. Supination: **a. Upward or palm upward**

145. Eversion: **f. Turning of the sole of the foot outward**

146. Inversion: **c. Turning of the sole of the foot inward**

147. Circumduction: **h. Forming of a cone with the arm or leg**

Where does the biceps muscle attach on the radius? d. The radial tuberosity

Which of these bones is not part of the pelvic girdle? b. Lumbar vertebrae

What is the formal term for the prominence commonly referred to as the elbow? a. Olecranon process

What is the formal term for the socket at the head of the femur? b. Acetabulum

Which is the largest and strongest tarsal bone? d. The calcaneus

Where are people who complain about their sacroiliac experiencing pain? a. Lower back

Sutures of the skull: a. Fibrous joint

Between carpals: d. Synovial joint

Knee joint: d. Synovial joint

Symphysis pubis: c. Cartilaginous joint - symphysis

Epiphyseal plate: b. Cartilaginous joint - synchondrosis

Intervertebral discs: c. Cartilaginous joint - symphysis

Following is how Figure 7-8, the synovial joint, should be labeled. 128, b. Periosteum; 129, e. Fibrous capsule; 130, a. Synovial (joint) cavity; 131, d. Articular cartilage; 132, c. Synovial membrane

What is the formal term for an immovable joint? b. A synarthrosis

What material or structure allows free movement in a joint? c. Synovial fluid

Choose an example of a ball-and-socket joint. a. The hip

Choose an example of a pivotal joint. a. Between the radius and the ulna

What is the term for the structure in the knee that divides the synovial joint into two separate compartments? d. Meniscus or articular disc

Flexion: b. Decrease in the angle between two bones

Extension: e. Increase in the angle between two bones

Abduction: g. Movement away from the midline of the body

Adduction: f. Movement toward the midline of the body

Rotation: i. Turning around an axis

Pronation: d. Downward or palm downward

Supination: a. Upward or palm upward

Eversion: ? Turning of the sole of the foot outward

Inversion: c. Turning of the sole of the foot inward

Circumduction: h. Forming of a cone with the arm or leg

Chapter **8**

Getting in Gear: The Muscles

Much of what we think of as "the body" centers around our muscles and what they can do, what we want them to do, and how tired we get trying to make them do it. With all that muscles do and are, it's hard to believe the word "muscle" is rooted in the Latin word *musculus*, which is a diminutive of the word for "mouse." Well, the muscle is a mouse that roars. Muscles make up most of the fleshy parts of the body and account for 30-50 percent of the body's weight (depending on sex with females being on the lower end). Layered over the skeleton (which we discuss in Chapter 7), they largely determine the body's form. More than 500 muscles are large enough to be seen by the unaided eye, and thousands more are visible only through a microscope.

There are three distinct types of muscle tissue, classified by their differing structure and control, but they share the following characteristics:

» **Contractility:** Ability to shorten

» **Extensibility:** Ability to stretch

» **Elasticity:** Ability to return to resting length

» **Excitability:** Ability to respond to stimulus

» **Conductivity:** Ability to create an electrical signal

Flexing Your Muscle Knowledge

REMEMBER

The study of muscles is called *myology* after the Greek word *mys*, which also means "mouse." Muscles perform a number of functions vital to maintaining life, including

>> **Movement:** Skeletal muscles (those attached to bones) convert chemical energy into mechanical work, producing movement ranging from finger tapping to a swift kick of a ball by contracting, or shortening. Reflex muscle reactions protect your fingers when you put them too close to a fire and startle you into watchfulness when an unexpected noise sounds. Many purposeful movements require several sets, or groups, of muscles to work in unison.

>> **Vital functions:** Without muscle activity, you die. Muscles are doing their job when your heart beats, when your blood vessels constrict, and when your intestines squeeze food along your digestive tract in *peristalsis*.

>> **Antigravity:** Perhaps that's overstating it, but muscles do make it possible for you to stand and move about in spite of gravity's ceaseless pull. Did your mother tell you to improve your posture? Just think how bad it would be without any muscles!

>> **Heat generation:** You shiver when you're cold and stamp your feet and jog in place when you need to warm up. That's because chemical reactions in muscles create heat as a byproduct, helping to maintain the body's temperature.

>> **Joint stability:** Muscles and their tendons aid the ligaments to reinforce joints.

>> **Supporting and protecting soft tissues:** The body wall and floor of the pelvic cavity support the internal organs and protect soft tissues from injury.

>> **Sphincters:** These muscles encircle openings to control swallowing, defecation, and urination.

As you may remember from studying tissues (see Chapter 5 for details), muscle cells — called *fibers* — are some of the longest in the body. Some muscle fibers contract rapidly, whereas others move at a leisurely pace. Generally speaking, however, the smaller the structure to be moved, the faster the muscle action. Exercise can increase the thickness of muscle fibers, but it doesn't make new fibers. Skeletal muscles have a rich vascular supply that dilates during exercise to give the working muscle the extra oxygen it needs to keep going.

Two processes are central to muscle development in the developing embryo: *myogenesis*, during which muscle tissue is formed, and *morphogenesis*, when the muscles form into internal organs. By the eighth week of gestation, a fetus is capable of coordinated movement.

REMEMBER

Following are some important muscle terms to know:

>> **Fascia:** Layers of dense connective tissue found between muscles; maintains a muscle's position relative to the other muscles around it

>> **Fiber:** An individual muscle cell

>> **Insertion:** The more movable attachment of a muscle

>> **Origin:** The immovable attachment of a muscle, or the point at which a muscle is anchored by a tendon to the bone

>> **Motor neuron:** A neuron that stimulates contraction of a muscle

>> **Myofibril:** Fibrils (fiber-like structures) within a muscle cell that contain the protein filaments actin and myosin that slide during contraction, shortening the fiber (or cell)

>> **Striated:** A characteristic of cardiac and skeletal muscle tissues; span the width of the cell

>> **Sarcoplasm:** The cellular cytoplasm in a muscle fiber; similarly the *sarcolemma* is the cell membrane

>> **Tendon:** Dense regular connective tissue made up of collagen, a fibrous protein that attaches muscles to bone; allows some muscles to apply their force at some distance from where a contraction actually takes place

>> **Aponeurosis:** Broad sheet of connective tissue that attaches muscles to other muscles (or sometimes bones)

>> **Tone, or tonus:** State of tension present to a degree at all times, even when the muscle is at rest

Complete the following practice questions to see how well you understand the basics of myology:

1 Which of the following is NOT a true statement?

a. Muscle fibers are some of the longest cells in the body.

b. Myofibrils within muscle cells contain protein filaments that slide during contraction.

c. Sphincters hold fibers together to form muscles.

d. Muscles create actions that both move the body and move things through it.

e. Posture is an expression of muscle action.

2 What happens during myogenesis and morphogenesis?

a. Muscles that aren't flexed regularly begin to lose tone during myogenesis and are destroyed entirely during morphogenesis.

b. Collagen is converted into muscle tissue during myogenesis and recycled by the body during morphogenesis.

c. Chemical energy is converted into mechanical work during myogenesis and morphogenesis.

d. Embryonic muscles form during myogenesis, and muscles form into internal organs during morphogenesis.

3 Why do your muscles shiver when you're cold?

    a. To prevent paralysis

    b. To generate heat

    c. To improve elasticity

    d. To increase excitability

    e. To maintain contractility

4 Name the cellular unit in muscle tissue.

    a. Filament

    b. Myofibril

    c. Fiber

    d. Motor cell

    e. Fasciculus

5 What defines a state of tension present to a degree at all times?

    a. Rigor

    b. Tonus

    c. Clovus

    d. Excitability

    e. Paralysis

6 It's possible to completely relax every muscle in the body.

    a. True

    b. False

7 Exercise forms new muscle fibers.

    a. True

    b. False

Muscle Classifications: Smooth, Cardiac, and Skeletal

REMEMBER

Muscle tissue is classified in three ways based on the tissue's function, shape, and structure:

>> **Smooth muscle tissue:** So-called because these spindle-shaped fibers don't have the striations typical of other kinds of muscle. This muscle tissue forms into sheets and makes up the walls of hollow organs such as the stomach, intestines, and bladder. The tissue's involuntary movements are relatively slow, so contractions last longer than those of other muscle tissue, and fatigue is rare. If arranged in a circle inside an organ, contraction constricts the cavity inside. If arranged lengthwise, contraction of smooth muscle tissue shortens the organ.

>> **Cardiac muscle tissue:** Found only in the heart, cardiac muscle fibers are cylindrical, striated, feature one central nucleus, and move through involuntary control. An electron microscope view of the tissue shows separate fibers tightly pressed against each other, forming cellular junctions called *intercalated discs* that look like tiny, dark-colored plates. These intercalated discs are gap junctions that help move an electrical impulse throughout the heart. For more on cardiac muscle, flip to Chapter 11.

>> **Skeletal muscle tissue:** This is the tissue that most people think of as muscle. It's the only muscle that can be voluntarily controlled via the somatic nervous system (SoNS). The long, striated, cylindrical fibers of skeletal muscle tissue contract quickly but tire just as fast. Skeletal muscle, which is also what's considered meat in animals, is 20 percent protein, 75 percent water, and 5 percent organic and inorganic materials. Each multinucleated fiber is encased in a thin, transparent cell membrane called a *sarcolemma* that receives and conducts stimuli. Packed into the cells are numerous tiny myofibrils, roughly 1 micron in diameter, that are suspended in the cell's sarcoplasm. Some skeletal muscle fibers have *muscle spindles* wrapped around them. These specialty sensory cells function in *proprioception*. They communicate with the brain to relay information on muscle actions which is translated into a sense of body position.

The following practice questions test your knowledge of muscle classifications:

8 What type of muscle tissue lacks striations?

a. Cardiac

b. Smooth

c. Skeletal

d. Fascia

e. Contracting

9 What does a sarcolemma do?

a. It receives and conducts stimuli for skeletal muscle tissue.

b. It provides a structure for moving electrical impulses through the heart.

c. It forms a wall for hollow organs.

d. It is made up of the filaments responsible for contraction.

e. It prevents muscle fatigue.

10 Which muscle type appears only in a single organ?

a. Contractile

b. Smooth

c. Cardiac

d. Skeletal

e. Fascia

11 Intercalated discs

a. anchor skeletal muscle fibers to one another.

b. play a role in moving electrical impulses through the heart.

c. are found only in the muscles of the back.

d. overlap to cause contraction.

e. contribute to tactile perception.

12–14 Identify the muscle tissue illustrated in Figure 8-1

12. _____   13. _____   14. _____

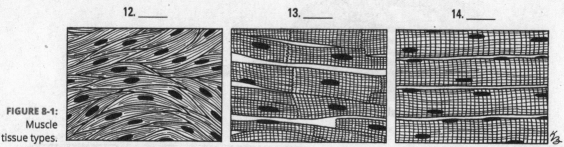

FIGURE 8-1:
Muscle
tissue types.

Illustration by Kathryn Born, MA

a. skeletal

b. smooth

c. cardiac

Contracting for a Contraction

Before we can explain how and why muscles do what they do, it's important that you understand the anatomy of how they're put together. Use Figure 8-2 as a visual guide as you read through the following sections.

Breaking down a muscle's anatomy and movement

We base this description of muscle on the most studied classification of muscle: skeletal. Each fiber contains hundreds, or even thousands, of myofibril strands made up of alternating filaments of the proteins *actin* (thin filaments) and *myosin* (thick filaments). Actin and myosin are what give skeletal muscles their striated appearance, with alternating dark and light bands. The dark bands are called *anisotropic*, or *A bands*. The light bands are called *isotropic*, or *I bands*. In the center of each I band is a line called the *Z line* that divides the myofibril into smaller units called *sarcomeres*. At the center of the A band is a less-dense region called the *H zone*. The H zone contains the *M line*, a fine filamentous structure that holds the thick myosin filaments in parallel arrangement. Table 8-1 summarizes the structure of a sarcomere.

For the muscle to contract, the sarcomere must shorten; that is, the Z lines get pulled closer together. This is accomplished by the interaction between actin and myosin. As the filaments slide past each other, the H zone is reduced or obliterated, pulling the Z lines closer together and reducing the I bands. (The A bands don't change.) Voilà! Contraction has occurred! This is explained in the sliding filament theory, which is illustrated in Figure 8-3.

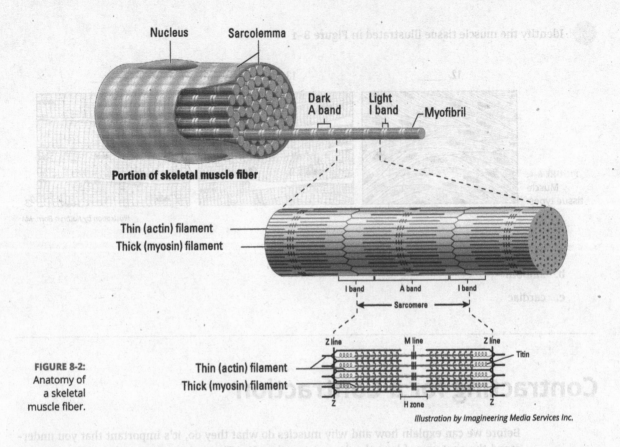

FIGURE 8-2:
Anatomy of a skeletal muscle fiber.

Illustration by Imagineering Media Services Inc.

Table 8-1 Structures of a Sarcomere

Structure	Description	Contents
Z-line	Thin line down the middle of the light band (I)	The protein that holds actin filaments in position, also attaches to myosin via titin
M-line	Thin line down the middle of the dark band (A)	Proteins that keep myosin filaments in line with each other
A-band	Dark block that creates striations	Myosin and any other proteins found within its length such as actin and titin
H-zone	Lighter block in the center of the dark band	Area of myosin that does not have actin overlapping it
I-band	Light band that creates striations	Actin filaments that don't overlap with myosin, the Z lines, and titin

The myosin filaments contain numerous heads that point away from the center of the sarcomere (see Figure 8-3a). The actin filaments contain binding sites for these myosin heads. Contraction begins when these heads grab onto the actin, forming a *cross bridge*, and perform *power strokes* (see Figure 8-3b). The first part of a power stroke requires no energy — cross bridges form and the myosin heads reflexively pull in. This alone, however, does not generate enough force to move a body part. To finish a power stroke, ATP comes in and binds to the myosin head, causing it to let go of the actin. Then, ATP is catalyzed and the energy is used to cock the heads back into their original position. (For a review of ATP, refer to Chapter 2.) The myosin heads are now able to perform another power stroke, this time grabbing a binding site further down on the actin filament. As long as the muscle is being stimulated and ATP is available, the *cross bridge cycling* can continue.

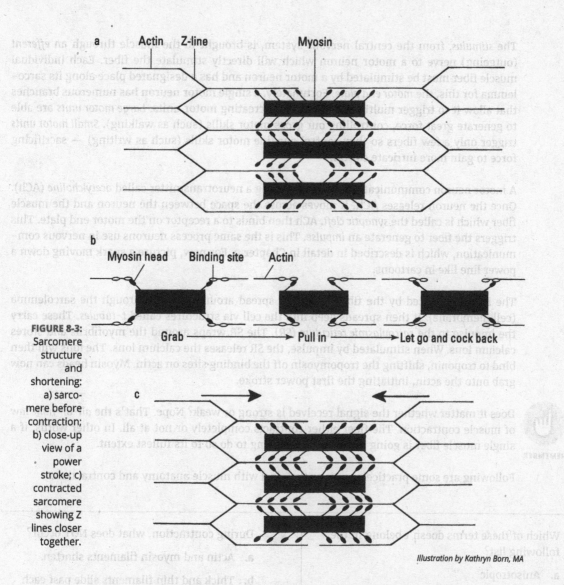

a

Actin   Z-line   Myosin

b

Myosin head   Binding site   Actin

Grab ──────→ Pull in ──────→ Let go and cock back

c

FIGURE 8-3:
Sarcomere
structure
and
shortening:
a) sarco-
mere before
contraction;
b) close-up
view of a
power
stroke; c)
contracted
sarcomere
showing Z
lines closer
together.

Illustration by Kathryn Born, MA

Because it requires no energy to form a cross bridge between a myosin head and the actin, the binding sites must remain covered until contraction is triggered by the associated motor neuron. This is achieved through a pair of proteins known as the *t-t complex*. *Tropomyosin* wraps around the actin filament, covering up the binding sites. *Troponin* pins the tropomyosin to the actin, ensuring it doesn't slip off. The final step of stimulation uncovers the binding sites.

Understanding what stimulates muscle contraction

After you know how muscles contract, you need to figure out what stimulates them to do so. We cover the details of the nervous system in Chapter 9, but here you can find out what's happening as an impulse stimulates a skeletal muscle.

The *stimulus*, from the central nervous system, is brought to the muscle through an *efferent* (outgoing) nerve to a motor neuron which will directly stimulate the fiber. Each individual muscle fiber must be stimulated by a motor neuron and has a designated place along its sarcolemma for this, the *motor end plate*. Fortunately, a single motor neuron has numerous branches that allow it to trigger multiple fibers at once, creating motor units. *Large motor units* are able to generate great force, controlling our gross motor skills (such as walking). *Small motor units* trigger only a few fibers so they control our fine motor skills (such as writing) — sacrificing force to gain more intricate control.

A motor neuron communicates with a fiber using a neurotransmitter called *acetylcholine* (ACh). Once the neuron releases ACh, it moves across the space between the neuron and the muscle fiber which is called the *synaptic cleft*. ACh then binds to a receptor on the motor end plate. This triggers the fiber to generate an *impulse*. This is the same process neurons use in nervous communication, which is described in detail in Chapter 9. For now, picture a spark moving down a power line like in cartoons.

The impulse created by the fiber is quickly spread around the cell through the sarcolemma (cell membrane). It then spreads deep into the cell via structures called *t-tubules*. These carry the impulse to the *sarcoplasmic reticulum (SR)*. The SR wraps around the myofibrils and stores calcium ions. When stimulated by impulse, the SR releases the calcium ions. The ions will then bind to troponin, shifting the tropomyosin off the binding sites on actin. Myosin heads can now grab onto the actin, initiating the first power stroke.

REMEMBER

Does it matter whether the signal received is strong or weak? Nope. That's the *all-or-none law* of muscle contraction. The fiber either contracts completely or not at all. In other words, if a single muscle fiber is going to contract, it's going to do so to its fullest extent.

Following are some practice questions that deal with muscle anatomy and contraction:

 15   Which of these terms doesn't belong in the following list?

a. Anisotropic

b. Actin

c. Myosin

d. Sarcolemma

e. Troponin

 16   During contraction, what does NOT occur?

a. Actin and myosin filaments shorten.

b. Thick and thin filaments slide past each other.

c. Muscle fibers shorten.

d. Myosin heads pull on actin filaments.

e. Troponin binds with calcium ions.

17 A weak stimulus causes a muscle fiber to contract only partway.

    a. True

    b. False

18 Which of the following steps of muscle contraction requires the breakdown of ATP?

    a. release of calcium ions

    b. uncovering of binding sites

    c. myosin heads pulling actin closer

    d. cross bridges breaking off from the binding sites

    e. myosin heads cocking back to their original position

19-23 Label the sarcomere illustrated in Figure 8-4

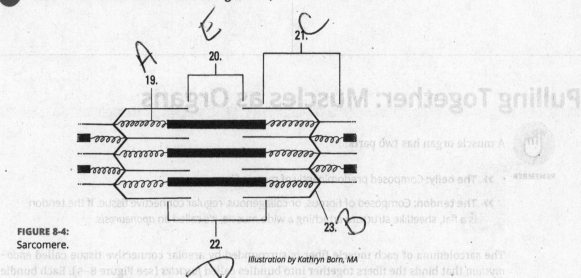

FIGURE 8-4:
Sarcomere.

Illustration by Kathryn Born, MA

    a. Titin

    b. Z lines

    c. I band

    d. A band

    e. H zone

24 Which statement best characterizes a large motor unit?

   a. Many neurons stimulate the same muscle, generating a large force.

   b. One neuron stimulates all the fibers of a muscle for precise control.

   c. It controls precise movements such as writing.

   d. Many neurons stimulate many fibers for a powerful contraction.

   e. One neuron stimulates many fibers for a powerful movement.

25 Outline the steps of contraction stimulus, ending with exposure of the binding sites.

_____

_____

_____

_____

_____

_____

Pulling Together: Muscles as Organs

REMEMBER

A muscle organ has two parts:

>> **The belly:** Composed predominantly of muscle fibers.

>> **The tendon:** Composed of fibrous, or collagenous, regular connective tissue. If the tendon is a flat, sheetlike structure attaching a wide muscle, it's called an *aponeurosis*.

The sarcolemma of each muscle fiber is surrounded by areolar connective tissue called *endomysium* that binds the fibers together into bundles called *fascicles* (see Figure 8-5). Each bundle is surrounded by areolar connective tissue called *perimysium*. All the fascicles together make up the belly of the muscle, which is surrounded by dense connective tissue called the *epimysium*. Blood vessels, lymphatic vessels, and nerves can be found wrapped in the epimysium as well as between the fascicles.

When actin and myosin overlap, the sarcomeres shorten pulling the ends of the myofibrils closer together. Since the myofibrils span the length of the fiber, the fiber shortens as a result. Similarly, the fibers run about the length of the muscle so the entire belly shortens, pulling on the tendon.

Cardiac and smooth muscle contract in the same manner but are organized into their organs differently. Cardiac muscle makes up the majority of the heart's walls. The cells, however, connect end-to-end, often to multiple cells (giving it its branched appearance). The myofibrils are still packed in parallel within the fibers so it is also striated in appearance. Cardiac muscle fibers usually have only one, centrally located nucleus. Smooth muscle is the least organized of the three. Like cardiac fibers, it has a single, central nucleus. The myofibrils, however, are not bundled in parallel, they crisscross the cell. As a result, smooth muscle is not striated, and instead of contracting by pulling ends together, it pulls in from all directions. This is why smooth muscle is found in the walls of hollow organs.

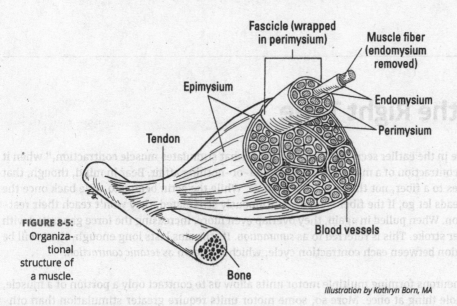

FIGURE 8-5: Organizational structure of a muscle.

Illustration by Kathryn Born, MA

 26-30 **Match the muscle structures with their descriptions.**

a. Perimysium

b. Aponeurosis

c. Epimysium

d. Fascicles

e. Endomysium

26 ___E___ Areolar tissue that surrounds fibers

27 ___D___ Bundles of muscle fibers

28 ___A___ Connective tissue that wraps muscle fibers into bundles

29 ___C___ Connective tissue that creates a functional muscle

30 ___B___ Flat, sheet-like tendon that serves as insertion for a large, flat muscle

 31 **Which of the following muscle tissues are striated?**

I. skeletal

II. cardiac

III. smooth

a. I only

b. III only

c. I and III

d. I and II

e. all of the above

32   While structurally different, all three muscle fiber types contract their myofibrils utilizing the sliding-filament model.

    a. True

    b. False

Assuming the Right Tone

As we note in the earlier section "Understanding what stimulates muscle contraction," when it comes to contraction of a muscle fiber, it's an all-or-nothing affair. Bear in mind, though, that this applies to a fiber, not the muscle as a whole. While the actin begins to slide back once the myosin heads let go, if the fiber is being continuously stimulated, they won't reach their resting position. When pulled in again, they overlap even more, increasing the force generated with each power stroke. This is referred to as *summation*. If stimulus lasts long enough, there will be no relaxation between each contraction cycle, which is known as *tetanic contraction*.

Multiple neurons forming multiple motor units allow us to contract only a portion of a muscle, or the whole thing at once. More so, some motor units require greater stimulation than others which allows us to control the force of contraction. Fibers that are easier to stimulate are triggered first. As the stimulus intensifies, more motor units are activated, which is referred to as *recruitment*. Recruitment ensures that muscle movements are smooth rather than jerky as the fibers *summate* to sustained contraction. Even when a muscle is at rest, some of its fibers maintain a level of sustained contraction called *tone*, or *tonus*. Without this, we wouldn't be able to stand or even hold our heads up!

REMEMBER

In physiology, a muscle contraction is referred to as a *muscle twitch*. A twitch is the fundamental unit of recordable muscular activity. The twitch consists of a single stimulus-contraction-relaxation sequence in a muscle fiber. The contraction is the period when the cross bridges are cycling. The relaxation phase occurs when cross bridges detach and the muscle tension decreases. Complete fatigue occurs when no more twitches can be elicited, even with increasing intensity of stimulation.

REMEMBER

There are two distinct types of muscle contraction:

>> **Isometric:** Occurs when a contracting muscle is unable to move a load, for example, trying to push a building to one side or doing a wall sit. It retains its original length but develops *tension.* No mechanical work is accomplished, and all energy involved is expended as heat. The muscles that control your posture maintain isometric contractions.

>> **Isotonic:** Occurs when the resistance offered by the load (or the gardening hoe or the cold can of soda) is less than the tension developed, thus shortening the muscle and resulting in mechanical work. There are two main types of isotonic contraction:

    • **Concentric contraction:** Occurs when the force generated by the muscle is less than the maximum tension. The muscle shortens with increased velocity. An example is the upward motion of a biceps curl (elbow flexion).

- **Eccentric contraction:** Occurs when the force of a load is greater than the maximum muscle tension, causing the muscle to elongate while performing work. This often occurs in conjunction with a concentric contraction on the opposite side. For example, the downward motion of a biceps curl: While the triceps perform the concentric contraction, the controlled motion demands an eccentric contraction from the biceps.

As such, muscles aren't independent, sole proprietors. Each muscle depends upon companions in a muscle group to assist in executing a particular movement. That's why muscles are categorized by their actions. The brain coordinates the following groups through the cerebellum.

>> **Prime movers:** Provide the major force for producing a specific movement. Just as it sounds, these muscles are the workhorses that produce movement.

>> **Antagonists:** These muscles exist in opposition to prime movers, regulating the motion by contracting and providing resistance.

>> **Fixators or fixation muscles:** These muscles serve to steady a part while other muscles execute movement. They don't actually take part in the movement itself.

>> **Synergists:** These muscles control movement of the proximal joints so that the prime movers can bring about movements of distal joints.

Flex your knowledge of muscle tone and function with these practice questions:

EXAMPLE

Q. Muscle contraction that is unable to move an object involves an action known as

   a. isometric.

   b. eccentric.

   c. isotonic.

   d. concentric.

A. The correct answer is isometric. When the tension leads to movement (actual work), it's isotonic. Eccentric contraction and concentric contraction are the two types of isotonic contraction.

---

 33   What do you call muscles that tend to counteract or slow an action?

   a. Antagonists

   b. Fixators

   c. Primary movers

   d. Agonists

   e. Synergists

 34   What creates a strong muscle contraction?

   a. Isotonic influences building up as mechanical work

   b. Isometric influences building up as tension

   c. Low frequency stimulatory nerve impulses

   d. High frequency stimulatory nerve impulses

35-38 Match the terms with their definitions.

a. Summation

b. Tone

c. Recruitment

d. Tetanus

35 __C__ As stimulation increases, more motor units are triggered to contract.

36 __A__ When twitches occur before the previous one ends, generating more force.

37 __D__ Stimulation occurs so frequently that no relaxation occurs at all.

38 __B__ Constant, sustained contraction of some fibers within a muscle.

Leveraging Muscular Power

Skeletal muscle power is nothing without lever action. The bone acts as a rigid bar, the joint is the fulcrum, and the muscle applies the force. Levers are divided into the *weight arm*, the area between the fulcrum and the weight; and the *power arm*, the area between the fulcrum and the force. When the power arm is longer than the weight arm, less force is required to lift the weight, but range, or distance, and speed are sacrificed. When the weight arm is longer than the power arm, the range of action and speed increase, but power is sacrificed. Therefore, 90 degrees is the optimum angle for a muscle to attach to a bone and apply the greatest force.

REMEMBER

Three classes of levers are at work in the body:

>> **First-class, or seesaw:** The fulcrum is located between the weight and the force being applied. An example is a nod of the head: The head-neck joint is the fulcrum, the head is the weight, and the muscles in the back of the neck apply the force.

>> **Second-class, or wheelbarrow:** The weight is located between the fulcrum and the point at which the force is applied. An example is standing on your tiptoes: The fulcrum is the joint between the toes and the foot, the weight is the body, and the muscles in the back of the leg at the heel bone apply the force.

>> **Third class, or removing a nail with a hammer:** The force is located between the weight and the fulcrum. An example is flexing your arm and showing off your biceps: The elbow joint is the fulcrum, the weight is the lower arm and hand, and the biceps insertion on the lower arm applies the force.

The power behind muscle contraction comes from the breakdown of ATP, which is generally made via cellular respiration. There are fibers, though, that can create ATP through lactic acid fermentation (see Figure 2-11 in Chapter 2). These fibers, referred to as *fast-twitch*, do not need oxygen and can thus respond very quickly, generating great force. However, this anaerobic process generates only two ATP per glucose whereas aerobic respiration produces 38. So these fast-twitch fibers cannot maintain contraction to near as long as their *slow-twitch* counterparts. Slow-twitch fibers are thus used for endurance, but they don't contain as many myofibrils, so they can't generate as much force. These fibers need room for *myoglobin*, which stores oxygen for respiration. All of our muscles contain both fiber types as well as intermediate fibers (which fall between slow- and fast-twitch in their characteristics). We can't rely on slow-twitch alone because they cycle more slowly and don't generate as much force. And we can't rely on fast-twitch alone because they tire out quickly, and the fermentation creates *lactic acid* as a byproduct.

WARNING

When lactic acid builds up, fibers can *fatigue:* They become unable to contract. This is distinct from a *cramp*, which is a sustained, involuntary contraction. While cramps and fatigue can occur simultaneously within a muscle, cramps come about from an electrolyte imbalance in the fluid surrounding the cell, not the accumulation of lactic acid. Got all that? Then try your hand at the following questions:

---

39 Which of the following in Figure 8-6 is a Class II lever?

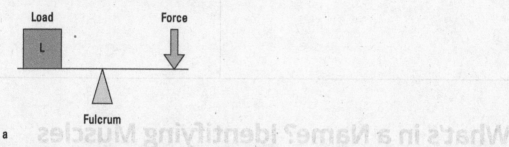

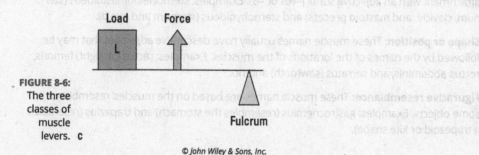

FIGURE 8-6:
The three classes of muscle levers.

© John Wiley & Sons, Inc.

40 Which muscles would provide the force in a Class III lever?

   a. Biceps brachii

   b. Pectoralis major

   c. Orbicularis oculi

   d. Rectus abdominus

   e. Sartorius

41 Which of the following would produce a wide range of movement with speed while sacrificing power?

   a. Power arm and weight arm of equal lengths

   b. Long weight arm, short power arm

   c. Long power arm, shorter weight arm

   d. Short power arm, shorter weight arm

What's in a Name? Identifying Muscles

REMEMBER

It may seem like a jumble of meaningless Latin at first, but muscle names follow a strict convention that names them according to one or more of the following:

» **Function:** These muscle names usually have a verb root and end in a suffix (*-or* or *-eus*), followed by the name of the affected structure. Example: levator scapulae (elevates the scapulae).

» **Compounding points of attachment:** These muscle names blend the origin and insertion attachment with an adjective suffix (*-eus* or *-is*). Examples: sternocleidomastoideus (sternum, clavicle, and mastoid process) and sternohyoideus (sternum and hyoid).

» **Shape or position:** These muscle names usually have descriptive adjectives that may be followed by the names of the locations of the muscles. Examples: rectus (straight) femoris, rectus abdominis, and serratus (sawtooth) anterior.

» **Figurative resemblance:** These muscle names are based on the muscles' resemblance to some objects. Examples: gastrocnemius (resembles the stomach) and trapezius (resembles a trapezoid or kite shape).

Check out Figure 8-7 and Table 8-2 for a rundown of prominent muscles in the body and key points to remember about each one.

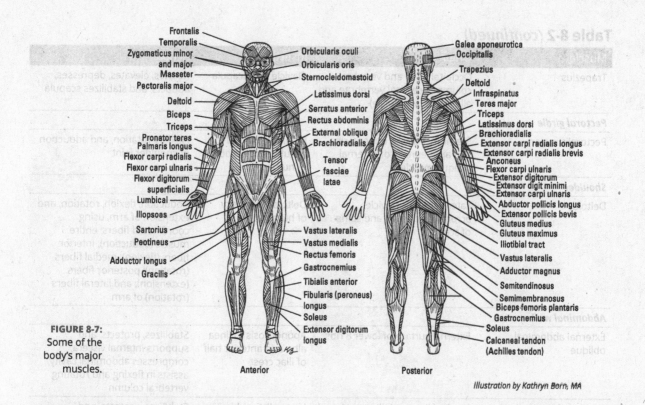

FIGURE 8-7:
Some of the
body's major
muscles.

Labels (Anterior):
Frontalis
Temporalis
Zygomaticus minor and major
Masseter
Pectoralis major
Deltoid
Biceps
Triceps
Pronator teres
Palmaris longus
Flexor carpi radialis
Flexor carpi ulnaris
Flexor digitorum superficialis
Lumbical
Iliopsoas
Sartorius
Peotineus
Adductor longus
Gracilis
Orbicularis oculi
Orbicularis oris
Sternocleidomastoid
Latissimus dorsi
Serratus anterior
Rectus abdominis
External oblique
Brachioradialis
Tensor fasciae latae
Vastus lateralis
Vastus medialis
Rectus femoris
Gastrocnemius
Tibialis anterior
Fibularis (peroneus) longus
Soleus
Extensor digitorum longus

Labels (Posterior):
Galea aponeurotica
Occipitalis
Trapezius
Deltoid
Infraspinatus
Teres major
Triceps
Latissimus dorsi
Brachioradialis
Extensor carpi radialis longus
Extensor carpi radialis brevis
Anconeus
Flexor carpi ulnaris
Extensor digitorum
Extensor digit minimi
Extensor carpi ulnaris
Abductor pollicis longus
Extensor pollicis bevis
Gluteus medius
Gluteus maximus
Iliotibial tract
Vastus lateralis
Adductor magnus
Semitendinosus
Semimembranosus
Biceps femoris plantaris
Gastrocnemius
Soleus
Calcaneal tendon (Achilles tendon)

Anterior   Posterior

Illustration by Kathryn Born, MA

Table 8-2   Muscles of the Body

Muscle	Origin	Insertion	Action
Head			
Frontalis	Galea aponeurotica	Eyebrow	Expression
Occipitalis	Occipital bone	Galea aponeurotica	Moves scalp forward and backward
Buccinator	Alveolar processes of mandible and maxillary bone	Orbicularis oris and skin at angle of mouth	Mastication
Orbicularis oculi	Encircles eye	Encircles orbit and extends within eyelid	Closes eye
Orbicularis oris	Encircles mouth	Skin surrounding mouth and at angle of mouth	Closes mouth
Masseter	Zygomatic arch	Mandible	Mastication
Temporalis	Temporal fossa	Coronoid process and ramus of mandible	Mastication
Zygomaticus	Zygomatic bone	Corner of mouth	Smiling
Neck			
Sternocleidomastoid	Manubrium of sternum and median portion of clavicle	Mastoid process of temporal bone	Rotation and flexion of neck vertebrae
Back			
Latissimus dorsi	Vertebral column	Humerus	Extends at shoulder joint

(continued)

Table 8-2 *(continued)*

Muscle	Origin	Insertion	Action
Trapezius	Occipital bone and vertebral column (7 cervical vertebrae and all thoracic vertebrae)	Clavicle and scapula	Rotates, elevates, depresses, adducts, and stabilizes scapula
Pectoral girdle			
Pectoralis major	Sternum, clavicle, upper 6 ribs, and aponeurosis of external oblique muscle	Medial margin of tertubercular groove of humerus	Flexion, rotation, and adduction of shoulder joint
Shoulder			
Deltoid	Lateral third of clavicle, acromion process, and spine of scapula	Deltoid tuberosity of humerus	Abduction, flexion, rotation, and extension of arm, using coordinated fibers: entire muscle (abduction); interior fibers (flexion); medial fibers (rotation); posterior fibers (extension); and lateral fibers (rotation) of arm
Abdominal wall			
External abdominal oblique	External surface of lower 8 ribs	Aponeurosis to linea alba and anterior half of iliac crest	Stabilizes, protects, and supports internal viscera, compresses abdominal cavity, assists in flexing and rotating vertebral column
Internal abdominal oblique	Inguinal ligament, iliac crest, and lumbodorsal fasciae	Linea alba, pubic crest, and last 3 ribs	Stabilizes, protects, and supports internal viscera, compresses abdominal cavity, assists in flexing and rotating vertebral column
Transversus abdominis	Inguinal ligament, iliac crest, lumbodorsal fasciae, and costal cartilage of last 6 ribs	Linea alba and pubic crest	Stabilizes, protects, and supports internal viscera, compresses abdominal cavity, assists in flexing and rotating vertebral column
Rectus abdominis	Pubic crest and symphysis pubis	Xiphoid process of sternum and costal cartilage of ribs 5–7	Stabilizes, protects, and supports internal viscera
Thorax			
Diaphragm	Xiphoid process of sternum, inner surface of lower 6 ribs, and lumbar vertebrae	Central tendon of diaphragm	Pulls central tendon downward, increasing size of thoracic cavity and causing inspiration; separates thoracic and abdominal cavities
External intercostals	Inferior border of rib above and costal cartilage	Superior border of rib below rib of origin	Elevates ribs, aiding inspiration
Internal intercostals	Superior border of rib below and costal cartilage	Inferior border of rib above rib of origin	Depresses ribs, aiding inspiration
Arm			
Biceps brachii	Long head: Tubercle above glenoid fossa, Short head: Coracoid process of scapula	Tuberosity of radius	Flexion at elbow joint

Muscle	Origin	Insertion	Action
Triceps brachii	Long head: Infraglenoid tubercle of scapula, Lateral head: Posterior surface of humerus above radial groove, Medial head: Posterior surface of humerus above radial groove	Olecranon process of ulna	Extension of elbow joint
Flexor carpi radialis	Medial epicondyle of humerus	2nd to 3rd metacarpals	Flexor of wrist, abducts hand
Flexor carpi ulnaris	Medial epicondyle of humerus and olecranon process	Pisiform bone and base of 5th metacarpal	Flexor of wrist, adducts hand
Supinator	Lateral epicondyle of humerus and proximal end of ulna	Proximal end of radius	Supinates forearm
Extensor carpi ulnaris	Lateral epicondyle of humerus	Base of 5th metacarpal	Extends and adducts wrist
Extensor carpi radialis longus	Lateral supracondylar ridge of humerus	Base of 2nd metacarpal	Extends and abducts wrist
Extensor carpi radialis brevis	Lateral epicondyle of humerus	Dorsal surface of 3rd metacarpal	Extends and abducts wrist, steadies wrist during finger flexion
Thigh, Anterior (Front)			
Rectus femoris (part of quadriceps)	Acetabulum of coxal bone	Tibial tuberosity via patella	Extends knee and flexes thigh at hip
Vastus lateralis (part of quadriceps), vastus medialis, vastus intermedialis (part of quadriceps)	Femur	Tibia	Extends knee
Sartorius	Anterior superior iliac spine	Proximal medial surface of tibia	Flexes at knee and laterally rotates thigh
Adductors	Pubis	Femur	Adduction and flexion at hip joint and lateral rotation of thigh
Gracilis	Pubis, symphysis pubis, and pubic arch	Medial surface of tibia	Adduction of thigh and rotation of leg
Thigh, Posterior (Back)			
Biceps femoris (part of hamstrings)	Long head: Ischial tuberosity, Short head: Linea aspera	Fibula and lateral condyle of tibia	Flexion at knee, extension of thigh
Semimembranosus (part of hamstrings)	Ischial tuberosity	Medial condyle of tibia	Flexion at knee, extension of thigh
Semitendinosus (part of hamstrings)	Ischial tuberosity	Proximal end of tibia	Flexion at knee, extension of thigh
Leg, Posterior			
Gastrocnemius	Medial and lateral condyles of femur	Calcaneus by Achilles (calcaneal) tendon	Flexion at knee and plantar flexion
Soleus	Posterior third of fibula and middle third of tibia	Calcaneus by Achilles tendon	Plantar flexion
Hip			
Gluteus maximus	Dorsal ilium, sacrum, and coccyx	Gluteal tuberosity of femur	Extends thigh and laterally rotates thigh

42 A muscle's name can be based on four things. Which one of those things was used to name these three muscles: the latissimus dorsi, the rectus abdominis, and the serratus anterior?

a. Shape

b. Attachment

c. Figurative resemblance

d. Function

43 The origin of the biceps brachii would best include which of the following?

a. Scapula

b. Clavicle

c. Fibula

d. Ulna

44-48 Match the origins and insertions for the following muscles.

a. The pubis and the femur

b. The femur and the calcaneus

c. The ilium and the tibia

d. The ischium and the tibia

e. The pubis and the tibia

44 _____ Semimembranosus

45 _____ Gracilis

46 _____ Sartorius

47 _____ Gastrocnemius

48 _____ Adductors

49-53 Match the muscles with their actions.

a. Rotation of scapula

b. Flexion of leg at knee joint

c. Extension at shoulder joint

d. Mastication

e. Flexion of forearm

49 _____ Semitendinosus

50 _____ Temporalis

51 _____ Biceps brachii

52 _____ Latissimus dorsi

53 _____ Trapezius

Match the muscles with their locations.

a. Head

b. Abdomen

c. Back

d. Neck

e. Thigh

54 _____ Latissimus dorsi

55 _____ Internal oblique

56 _____ Quadriceps

57 _____ Masseter

58 _____ Sternocleidomastoid

 What "two-headed" muscle is found in both the arm and the leg?

a. Quadriceps

b. Triceps

c. Biceps

d. Abdominals

e. Gluteus

 This muscle divides the thoracic cavity from the abdominal cavity.

a. Diaphragm

b. External oblique

c. Transversus abdominis

d. Rectus abdominus

e. Internal oblique

61 Which of the following is *not* one of the muscles referred to as hamstrings?

a. Biceps femoris

b. Gracilis

c. Semimembranosus

d. Semitendinosus

Answers to Questions on Muscles

The following are answers to the practice questions presented in this chapter.

(1) Which of the following is *not* a true statement? **c. Sphincters hold fibers together to form muscles.** Sphincters are rings of muscle fibers (cells) used as a control mechanism.

(2) What happens during myogenesis and morphogenesis? **d. Embryonic muscles form during myogenesis, and muscles form into internal organs during morphogenesis.**

(3) Why do your muscles shiver when you're cold? **b. To generate heat.** Chemical reactions in muscles result in heat, which is also why you stamp your feet and jog in place when you're cold.

(4) Name the cellular unit in muscle tissue. **c. Fiber.** When it comes to muscle, fibers and cells are the same thing.

(5) What defines a state of tension present to a degree at all times? **b. Tonus.** That's the elusive muscle "tone" for the flabby amongst us.

(6) It's possible to completely relax every muscle in the body. **b. False.** If every muscle in the body were to relax, the heart would stop beating and food would stop moving through the digestive system.

(7) Exercise forms new muscle fibers. **b. False.** Exercise can't form new fibers, it stimulates protein synthesis within the cells, causing them to enlarge.

(8) What type of muscle tissue lacks cross-striations? **b. Smooth.** Fascia is not muscle tissue and contracting muscle tissue is a made-up term.

(9) What does a sarcolemma do? **a. It receives and conducts stimuli for skeletal muscle tissue.** Other answers relate to other types of muscles, so if you associate "skeletal" with "sarcolemma," you'll stay on the right track.

(10) Which muscle type appears only in a single organ? **c. Cardiac.** And that sole organ is the heart.

(11) Intercalated discs **b. play a role in moving electrical impulses through the heart.** There's evidence that these structures help keep the heart synchronized.

(12) **b. Smooth muscle**

(13) **c. Cardiac muscle**

(14) **a. Skeletal muscle**

(15) Which of these terms doesn't belong in the following list? **d. Sarcolemma.** This is the cell membrane encasing the myofibrils. All the other answer options refer to various protein structures.

(16) During contraction, what does *not* occur? **a. Actin and myosin filaments shorten.** Quite the opposite: Muscle fibers shorten, actin and myosin merely overlap.

(17) A weak stimulus causes a muscle fiber to contract only partway. **b. False.** Muscle contractions are all-or-nothing; there's no such thing as a partial contraction.

18. Which of the following steps of muscle contraction requires the breakdown of ATP? **e. myosin heads cocking back to their original position.** While ATP must be present to cause the myosin heads to let go, the energy it contains is not used until the heads cock back.

19. **a. Titin**

20. **e. H zone**

21. **c. I band**

22. **d. A band**

23. **b. Z line**

24. Which statement best characterizes a large motor unit? **e. One neuron stimulate many fibers for a powerful movement.**

25. The steps should go something like this:

    Neuron releases ACh

    ACh spreads across the synaptic cleft

    ACh binds to the motor end plate

    Muscle fiber generates impulse

    Impulse spread through sarcolemma then through the t-tubules

    Impulse stimulates the SR which releases calcium ions

    Calcium ions bind to troponin

    Tropomysoin (or t-t complex) slides around, uncovering the binding sites

26. Areolar tissue that surrounds fibers: **e. Endomysium.**

27. Bundles of muscle fibers: **d. Fascicles.**

28. Connective tissue that wraps muscle fibers into bundles: **a. Perimysium.**

29. Connective tissue that creates a functional muscle **c. Epimysium.**

30. Flat, sheet-like tendon that serves as insertion for a large, flat muscle: **b. Aponeurosis.**

31. Which of the following muscle tissues are striated? **d. I and II:** While all three tissues have visible, length-wise stripes in the fibers, the striations are the stripes that run the length.

32. While structurally different, all three muscle fiber types contract their myofibrils utilizing the sliding-filament model. **a. True:** The difference is how the myofibrils are arranged within a cell.

33. What do you call muscles that tend to counteract or slow an action? **a. antagonists**

    TIP

    The muscles are against the action, so think of them as antagonistic.

(34) What creates a strong muscle contraction? **d. High frequency stimulatory nerve impulses.** The first two answers relate to types of muscle contractions, not strength.

(35) As stimulation increases, more motor units are triggered to contract: **c. Recruitment.**

(36) When twitches occur before the previous one ends, generating more force: **a. Summation.**

(37) Stimulation occurs so frequently that no relaxation occurs at all: **d. Tetanus.**

(38) Constant, sustained contraction of some fibers within a muscle: **b. Tone.**

(39) Which of the following in Figure 8-6 is a Class II lever? **b. Pectoralis major.** With Class II, remember to look for wheelbarrow-like leverage.

(40) Which muscles would provide the force in a Class III lever? **a. Biceps brachii.** It's the Popeye weight-lifting class, after all.

(41) Which of the following would produce a wide range of movement with speed while sacrificing power? **b. Long weight arm, short power arm.** The longer the weight arm, the greater the range of action and speed *but* the less power there is.

(42) A muscle's name can be based on four things. Which one of those things was used to name these three muscles: the latissimus dorsi, the rectus abdominis, and the serratus anterior? **a. Shape.** It's a matter of memorizing a bit of Latin: *latissimus* stems from the Latin word for "wide," *rectus* from the Latin word for "straight," and *serratus* from the Latin word for "notched" or "scalloped."

(43) In humans, the origin of the biceps brachii would best include which of the following? **a. Scapula.** After you know that your biceps brachii are in your upper arm, you just have to remember the name of the origin bone. The Latin meaning of *scapulae* is, literally, "shoulder blades," so perhaps a better memory tool would be that the scapula looks vaguely like a spatula.

(44) Semimembranosus: **d. The ischium and the tibia.**

(45) Gracilis: **e. The pubis and the tibia.**

(46) Sartorius: **c. The ilium and the tibia.**

(47) Gastrocnemius: **b. The femur and the calcaneus.**

(48) Adductors: **a. The pubis and the femur.**

(49) Semitendinosus: **b. Flexion of leg at knee joint.**

(50) Temporalis: **d. Mastication**

(51) Biceps brachii: **e. Flexion of forearm.**

(52) Latissimus dorsi: **c. Extension at shoulder joint.**

(53) Trapezius: **a. Rotation of scapula.**

(54) Latissimus dorsi: **c. Back.**

(55) Internal oblique: **b. Abdomen.**

(56) Quadriceps: **e. Thigh.**

(57) Masseter: **a. Head.**

(58) Sternocleidomastoid: **d. Neck.**

(59) What "two-headed" muscle is found in both the arm and the leg? **c. Biceps.** Your biggest clue: the "bi" part of biceps. Although people usually think of the biceps brachii in the arm, you can't forget about the biceps femoris at the back of the thigh.

(60) This muscle divides the thoracic cavity from the abdominal cavity. **a. Diaphragm.** And without it, you couldn't breathe.

(61) Which of the following is not one of the muscles referred to as hamstrings? **b. Gracilis.** The other three answer options all are listed as hamstring muscles.

3

Mission Control: All Systems Go

Chapter **9**

Feeling Jumpy: The Nervous System

Throughout this book, you look at the human body from head to toe, exploring how it collects and distributes the molecules it needs to grow and thrive, how it reproduces itself, and even how it gets rid of life's nastier byproducts. In this chapter, however, you look at the living computer that choreographs the whole show, the one system that contributes the most to making us who we are as humans: the nervous system.

In this chapter, you get a feel for how the nervous system is put together. You practice identifying the parts and functions of nerves and the brain itself as well as the structure and activities of the central, peripheral, and autonomic systems. In addition, we touch on the sensory organs that bring information into the human body.

Our Motherboard

The nervous system is the communications network that goes into nearly every part of the body, innervating your muscles, pricking your pain sensors, and letting you reach beyond yourself into the larger world. More than 80 major nerves make up this intricate network, with billions of *neurons* (individual nervous cells). It's through this complex network that you respond both to external and internal stimuli, demonstrating a characteristic called *irritability* (the capacity to respond to stimuli, not the tendency to yell at annoying people).

Put another way, there are three functions of the human nervous system as a whole:

>> **Sensation:** The ability to generate nerve impulses in response to changes in the external and internal environments (this also can be referred to as *perception*)

>> **Coordination:** The ability to receive, sort, and direct those signals to channels for response (this also can be referred to as *integration*)

>> **Reaction:** The ability to translate orders to the *effectors* (muscles and glands), which carry out the body's response to the stimulus

The nervous system has two divisions — the *central* and *peripheral nervous systems*. The central nervous system (CNS) includes the brain and spinal cord and is the body's integrator. Everything else falls into the peripheral nervous system. This is split into *sensory* and *motor divisions*. The sensory, or *afferent*, system responds to stimuli and communicates it to the CNS. If a response is warranted, the message is sent to the motor, or *efferent*, system, leading to the body's response to the stimulus. The motor system is further divided into the *somatic nervous system* which carries out voluntary commands and the *autonomic nervous system* which carries out all the automatic functions that we don't have to think about.

Before trying to study the system as a whole, it's best to break it down into building blocks first: neurons, nerves, impulses, and synapses.

Building from Basics: Neurons, Nerves, and Glial Cells

The nervous system is comprised mostly of neural tissue along with connective tissues for structure and to house blood and lymphatic vessels. There are two types cells found in neural tissue: *neurons* and *glial cells (neuroglia)*. *Nerves* are not cells, they are an organizational structure that bundles together extensions of the neurons called axons.

Neurons

The nervous cell responsible for the communication at the core of the nervous system's functions is the neuron. Its properties include that marvelous *irritability* that we speak of in the chapter introduction as well as *conductivity*, otherwise known as the ability to transmit a *nerve impulse*.

The central part of a neuron is the *cell body*, or *soma*, that contains a large nucleus with one or more nucleoli, the organelles common to all other cells, and *Nissl bodies*. (See Chapter 3 for an overview of a cell's primary parts.) The Nissl bodies are also found in the dendrites and are specialized pieces of rough ER thought to make proteins for the cell itself. The ER found in the cell body manufactures *neurotransmitters* that the neuron uses to communicate with other cells. Extending from the cell body are numerous cytoplasmic projections, containing specialized fibrils, or *neurofibrillae*, forming *dendrites* and *axons*. Dendrites conduct impulses to the cell body while *axons* conduct impulses away from it (see Figure 9-1). While a neuron can have numerous dendrites, it has only one axon; however, each axon can have many branches called *axon collaterals*, enabling communication with many target cells. The point of attachment on the cell body is called the *axon hillock*. The opposite end of the axon enlarges into a *synaptic knob* or *bouton*.

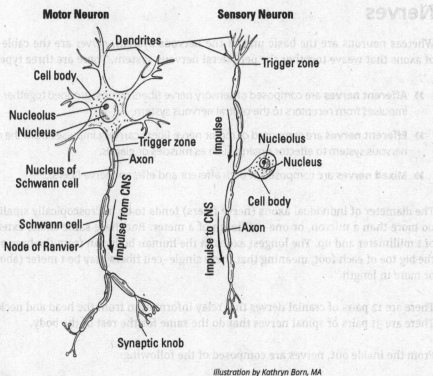

FIGURE 9-1: The motor neuron on the left and the sensory neuron on the right show the cell structures and the paths of impulses.

Illustration by Kathryn Born, MA

REMEMBER

There are four types of neurons, as follows:

>> **Multipolar:** Multiple dendrites separated from the axon by the cell body. Most motor, or efferent, neurons are of this type, connecting the CNS to its effectors — the structures that will carry out the effect. (Get it?)

>> **Unipolar:** Both the axon and dendrites come from the same cytoplasmic extension, meaning the dendrites connect directly to the axon rather than through the cell body. Most sensory, or afferent, neurons are unipolar carrying information about stimuli to the CNS.

>> **Bipolar:** A single dendrite and single axon (though they both have branches) meet the cell body on opposite sides. These are found almost exclusively in the CNS, the only exception being specialized sensory neurons from the eyes, nose, and ears.

>> **Anaxonic:** Axon cannot be distinguished from dendrites. These neurons are found in the brain, but their function is not well understood. They are thought to regulate motor neuron networks and may play a role in learning.

TIP

Here are a couple of handy memory devices: <u>A</u>fferent connections <u>a</u>rrive, and <u>e</u>fferent connections <u>e</u>xit. <u>D</u>endrites <u>d</u>eliver impulses while <u>a</u>xons send them <u>a</u>way.

Nerves

REMEMBER

Whereas neurons are the basic unit of the nervous system, *nerves* are the cable-like bundles of axons that weave together the peripheral nervous system. There are three types of nerves:

>> **Afferent nerves** are composed of sensory nerve fibers (axons) grouped together to carry impulses from receptors to the central nervous system.

>> **Efferent nerves** are composed of motor nerve fibers carrying impulses from the central nervous system to effector organs, such as muscles or glands.

>> **Mixed nerves** are composed of both afferent and efferent nerve fibers.

The diameter of individual axons (nerve fibers) tends to be microscopically small — many are no more than a micron, or one-millionth of a meter. But these same axons extend to lengths of 1 millimeter and up. The longest axons in the human body run from the base of the spine to the big toe of each foot, meaning that these single-cell fibers may be 1 meter (about three feet) or more in length.

There are 12 pairs of cranial nerves that relay information from the head and neck to the brain. There are 31 pairs of spinal nerves that do the same for the rest of the body.

From the inside out, nerves are composed of the following:

>> **Axon:** The impulse-conducting process of a neuron, also known as a fiber.

>> **Myelin sheath:** An insulating envelope that protects the axon and speeds up impulse transmission. (This process is discussed in detail in the "Impulses" section later in this chapter.)

>> **Neurolemma (or neurilemma):** A thin membrane present in many peripheral nerves that surrounds the myelin sheath and its enclosed axon.

>> **Endoneurium:** Loose, or *areolar*, connective tissue surrounding the length of individual fibers.

>> **Fasciculi:** Bundles of fibers within a nerve.

>> **Perineurium:** The same kind of connective tissue as endoneurium; bundles fascicles together.

>> **Epineurium:** The same kind of connective tissue as endoneurium and perineurium; surrounds all the bundles, creating the nerve.

Nerves organize the transport of signals both to and from the brain. Think about traveling across the country by car. Which is more efficient: taking an interstate or city streets? Nerves are the interstates of our nervous system's roads. Since nerves contain only the axons of peripheral neurons, a different structure organizes the cell bodies (and their dendrites). This structure, called a *ganglion*, bundles the cell bodies of related neurons (both afferent and efferent). These are found as the cranial and spinal nerves enter the CNS. The ganglia connect with each other forming a *plexus* that controls related body functions. For example, the solar plexus controls contractions throughout the digestive tract.

Glial cells

Much has been learned about glial cells in the past 30 years. At first, scientists thought that glial cells merely held neurons in their proper position, hence their name. (*Glia* is Greek for glue.) Until recently, they had been largely ignored by researchers, assuming they provided only structural support. However, we have learned that these "other nervous cells" have some pretty important functions. Table 9-1 describes the types of glial cells and their functions.

Table 9-1   Cranial Nerves

Glial Cell	Description	Function
Astrocyte	Star-shaped, found between neurons and blood vessels in CNS	Form blood brain barrier, takes up excess neurotransmitters and ions, damage repair, create new neuron connections
Ependyma	Often ciliated, forms simple cuboidal epithelium in CNS	Creates a filter controlling which molecules can move in and out of the CSF and the nervous tissue
Microglia	Small cell with tiny branches in CNS	Phagocytosis of debris, waste, and pathogens
Oligodendrocyte	Rounded cell with extensions that wrap around axons in the CNS	Form myelin sheaths, produce neurotrophic factors (growth and development of neurons)
Schwann cell	Small cell with extensions that wrap around axons in the PNS	Form myelin sheaths
Satellite cell	Flat cell surrounding cell bodies in ganglia of PNS	Protection, regulate nutrients

As you can see from Table 9-1, glial cells do far more than just being the glue that holds neurons together! Now let's see if you can answer the following questions about glial cells and the other cells and structures of the nervous system:

1-5 Fill in the blanks.

The communicating cells of the nervous system, or 1._____ have two essential properties. First is 2._____, the tendency to respond to a stimulus. Second is the ability to generate electricity, called 3._____. This flow of electricity is known as 4._____ but cannot be used to communicate with other neurons. For this, 5._____ are used.

6 What are the three primary functions of the nervous system?

a. Motion, circulation, conceptual thought

b. Action, conduction, filtration

c. Perception, integration, action

d. Circulation, coordination, response

7 Which structure is found at the end of an axon?

a. Soma

b. Dendrite

c. Nissl body

d. Ganglion

e. Synaptic knob

8 Which part of a neuron is responsible for conducting electrical signals toward the cell body?

a. axon

b. soma

c. bouton

d. dendrite

e. hillock

9 Which membrane surrounds an axon (nerve fiber)?

    a. Sarcolemma

    b. Neurolemma

    c. Perineurium

    d. Fasiculus

    e. Epineurium

 10–15 Match the term to its description.

    a. Cells that form and preserve myelin sheaths in PNS

    b. Cells that form and preserve myelin sheaths in CNS

    c. Cell that provide nutrients for bundled cell bodies

    d. Cells that are phagocytic

    e. Cells that form linings in the CNS

    f. Cells that contribute to the repair process of the central nervous system

    10 _____ Astrocytes

    11 _____ Microglia

    12 _____ Oligodendrocytes

    13 _____ Schwann cell

    14 _____ Satellite cell

    15 _____ Ependyma

 16 How many dendrites and axons are there on a multipolar neuron?

    a. Many dendrites and one axon

    b. Three dendrites and three axons

    c. Four or more dendrites and two axons

    d. A single dendrite and a single axon

    e. Numerous dendrites and numerous axons

17 A synapse between neurons is best described as the transmission of

    a. a continuous impulse.

    b. an impulse through chemical and physical changes.

    c. an impulse through a physical change.

    d. an impulse through a chemical change.

18 The connective tissue that surrounds a bundle of fibers in a nerve is known as the

a. Neurolemma

b. Perineurium

c. Endoneurium

d. Fasiculus

e. Epineurium

19 True or False: Nervous tissue is essentially neurons connected end-to-end with very little matrix.

20 Which statement best describes how neurons communicate?

a. Axons generate impulse which is passed to dendrites through the cell body.

b. Axons receive a stimulus that is passed to the cell body, triggering the dendrite to release a chemical.

c. Cell bodies respond to a stimulus by generating an electrical signal to pass on to dendrites and axons, where the former will pass it on chemically and the latter will do so electrically.

d. In response to stimulus, the dendrites generate an electrical signal that is passed to the cell body. The axon will then spread the impulse down its length releasing a chemical at the end.

Feeling Impulsive?

Neurons communicate using both electric and chemical signals. The electrical signal comes from the movement of ions — much like the flow of electrons we know as electricity. The chemical signal comes in the form of *neurotransmitters*. The entire communication process occurs in three phases: stimulus of the dendrites, impulse through the axon, and synaptic transmission from the synaptic knob.

Along the neuron

Before diving in to the communication process, you must first understand that a neuron is a *polarized* cell. Compared to its surroundings, the inside of the cell is more negatively charged — to the tune of −70mV. This is due to the high concentration of *anions* (negatively charged molecules) held there, which the neuron maintains when at rest, hence the term *resting potential*. Additionally, the axon has a high concentration of potassium ions (K⁺) inside and a high concentration of sodium ions (Na⁺) outside.

TIP

Don't let the positive charge of potassium ions throw you off. There are still numerous organic anions that counteract the K⁺ inside of the cell, creating the resting potential of −70mV.

Stimulus

The stimulus of a neuron generally happens in one of two ways. Most commonly, it is done by a neurotransmitter, a chemical messenger released from another neuron. The neurotransmitter binds to receptors on the dendrites, which triggers ion channels to open. The other option occurs only in sensory receptor neurons. Here, the designated stimulus causes the ion channels to open — for instance, heat in a hot thermoreceptor (but not cold!) or touch in a mechanoreceptor. Regardless of how the channels are triggered to open, positive ions rush into the cell (it's −70mV in the cell, after all). This causes the *depolarization* of the cell body as it becomes more positive (but overall still negatively charged). This initial depolarization, however, does not mean the neuron will proceed in communicating the signal. For that to occur, enough positive ions must come in to depolarize the cell body, where it meets the axon, to the threshold potential, which is −55mV (on average). If threshold is not reached, the ions are pumped out of the cell body and that's it.

REMEMBER

A single stimulus is generally not enough to warrant the neuron sending a message (just think of how exhausting it would be if we bombarded the brain with random, useless information). A larger (or repeated) stimulus will open more channels, causing more positive ions to rush in, and threshold is more likely to be reached. This causes the *all-or-none response* of a neuron. It either sends a message or it doesn't. Value of sensation ("this is warm" versus "whoa that's really hot!") is imparted by the brain based on how many neurons are sending the same message.

Impulse

The axon hillock is the *trigger zone* for impulse — the trigger being the reaching of threshold potential. This section of the axon has numerous *voltage-gated* sodium channels that pop open when the cell reaches −55mV. This is the beginning of the first *action potential*. A single action potential occurs in the following steps:

1. Sodium gates open.

2. Na⁺ rushes in, depolarizing that section of the axon.

3. This continues until enough Na⁺ comes in to trigger the voltage gated channels to close, which is +35mV on average.

4. Reaching 35mV also triggers potassium gates to open and K⁺ flows out.

5. The axon segment is now repolarized to the resting potential (−70mV).

6. Too many K⁺ leave causing *hyperpolarization* of the cells.

7. The hyperpolarization kick-starts the *sodium-potassium pumps* to move the ions against their concentration gradients to their original locations.

The Na/K pumps are the only step of action potential that requires ATP. Until all the ions have been replaced, the axon segment is in a *refractory period*, meaning it is unable to respond to another stimulus.

An action potential only carries us down one small segment of the axon — nowhere near its entire length. This is where *impulse*, or the sequential generation of new action potentials, comes in. When Na⁺ first rushes in, some of them flow down the axon increasing the charge within. Once enough slide over to hit −55mV, a new set of sodium gates open, generating the next action potential. This process repeats all the way to the synaptic knob while the individual segments continue to proceed through the process of action potential.

Remember in elementary school learning how to sing in the round? You were split into groups and sang "Row, row, row your boat," but each group started at a different time. This is how impulse works. If you follow the impulse, all you hear is "row, row, row" over and over again until the synaptic knob is reached. But if you only listen to the first group, you hear the entire song — the action potential.

While there is no scientific consensus on how often a neuron can fire, we do have a method to speed up the process: *saltatory conduction*. Many neurons have beads of *myelin* along the length of the axon. These sheets of lipids are created by glial cells and are wrapped around, leaving small sections of axon exposed called the *nodes of Ranvier*. When Na⁺ enters the axon they continue to flow down under the insulation of the myelin. A new action potential can only be initiated at the nodes (because any Na⁺ gates would be covered up). Thus, a myelinated neuron needs far fewer action potentials to span the length of the axon, saving energy (less ion pumping for recovery) and increasing signal speed up to 100 times.

Across the synapse

Unfortunately, most neurons cannot pass on an electrical signal to another cell, so our communication process is not yet complete. Instead, the electrical signal must be converted into a chemical one; this occurs at the synaptic knob.

When impulse reaches the end of the axon, the final action potential brings in an influx of Na⁺ as before. The immediate area depolarizes, but instead of opening sodium gates, calcium gates are opened instead (also triggered by reaching −55mV). Calcium ions rush into the synaptic knob where neurotransmitters are stored in vesicles. The calcium ions shove the vesicles to the cell membrane and exocytosis occurs (see Chapter 3), releasing the neurotransmitters into the *synapse* (the small space between the cells; they aren't physically connected). The *pre-synaptic* (sending) neuron has now completed its task and the *post-synaptic* (receiving) cell now repeats the process.

REMEMBER Neurotransmitters can be excitatory or inhibitory. The previous explanation assumes that the stimulating neurotransmitters are excitatory, opening channels for positive ions to enter the dendrites. Some neurotransmitters, however, are inhibitory, either closing channels or opening ones for negative ions to rush in. If a neuron had been repeatedly firing but was then stimulated by enough inhibitory neurotransmitters, it would no longer reach threshold and would stop its communication.

Now let's get your neurons firing to answer some questions.

 Match the term to its description.

a. A differential of charge from outside of the cell to the inside

b. Negatively charged molecule concentrated inside the cell

c. Threshold of excitation determines ability to respond

d. A drop in charge inside the cell past resting potential

e. An increase in charge inside of the cell

21 _____ All-or-none response

22 _____ Hyperpolarization

23 _____ Anion

24 _____ Polarization

25 _____ Depolarization

 26 Place the steps of action potential in order from 1 to 11:

8 _____ Potassium gates close

10 _____ Pumps restore polarization

3 _____ Cell depolarizes

2 _____ Sodium rushes in

6 _____ Potassium flows out

1 _____ Sodium gates open

4 _____ Sodium gates close

5 _____ Potassium gates open

11 _____ Ions return to initial concentrations

7 _____ Cell repolarizes

9 _____ Cell is hyperpolarized

 27 What is the role of Ca²⁺ ions in nervous communication?

a. Repolarize the cell during stimulation

b. Exits the cell to depolarize it during impulse transmission

c. Enters the cell to depolarize it during action potential

d. Repolarize the cell to activate the synaptic knob

e. Enters the cell to trigger the release of neurotransmitters

 28 True or False: Neurotransmitters bind to receptors on the dendrite opening channels for positive ions to rush in and depolarize the cell.

29   Explain the different between impulse and action potential.

_____

_____

_____

_____

Minding the Central Nervous System

Together, the brain and the spinal cord make up the central nervous system. You get the scoop on both parts in the following sections.

The spinal cord

REMEMBER

The spinal cord has two primary functions: It conducts nerve impulses to and from the brain, and it processes sensory information to produce a spinal reflex without initiating input from the higher brain centers.

The spinal cord begins at the brainstem and extends down into the lumbar region of the vertebral column. In an adult, the 18-inch spinal cord ends between the first and second lumbar vertebrae, roughly where the last ribs attach. Its tapered end is called the *conus medullaris*. The cord continues as separate strands below that point and is referred to as the *cauda equina* (horse tail). A thread of fibrous tissue called the *filum terminale* extends to the base of the *coccyx* (tailbone).

The spinal cord is an oval-shaped cylinder with two grooves running its length — the shallow *posterior median sulcus* at the back and the deeper *anterior median fissure* on the front. Also packed into the spinal cavity are the *meninges*, cerebrospinal fluid, a cushion of fat, and various blood vessels.

Three membranes called *meninges* envelop the central nervous system (brain and spinal cord), separating it from the bony cavities.

>> **The dura mater:** The outer layer, the *dura mater*, is the hardest, toughest, and most fibrous layer and is composed of white collagenous and yellow elastic fibers. It also contains numerous nerves, blood vessels, and lymphatic vessels.

>> **The arachnoid:** The middle membrane, the *arachnoid*, forms a weblike layer just inside the dura mater and does not contain blood vessels.

>> **The pia mater:** A thin inner membrane, the *pia mater*, lies along the surface of the central nervous system and like the dura, contains many blood vessels and nerves.

Between the arachnoid and the pia is the *subarachnoid space* which contains the cerebrospinal fluid (CSF) secreted by structures called *choroid plexuses*. The CSF flows through this space and circulates through the brain's ventricles, carrying nutrients pulled from the blood stream and wastes to be returned to it or into the lymphatic vessels.

Two types of solid material make up the inside of the cord, which you can see in Figure 9-2: *gray matter* (which is indeed grayish in color) containing unmyelinated neurons, dendrites, cell bodies, and neuroglia; and *white matter (funiculus)*, so-called because of the whitish tint of its myelinated nerve fibers. At the cord's midsection is a small *central canal* filled with CSF and surrounded first by gray matter in the shape of the letter H and then by white matter, which fills in the areas around the H pattern. The top, or skinny, legs of the H are the *posterior horns* and the bottom, thicker sections are the *anterior horns*. The cross bar is the *gray commissure* which houses the central canal.

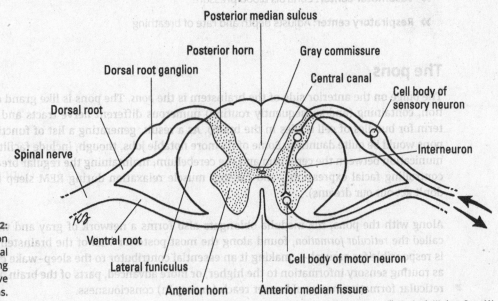

FIGURE 9-2: A cross-section of the spinal cord, showing spinal nerve connections.

Illustration by Kathryn Born, MA

The white matter consists of thousands of myelinated nerve fibers arranged in three *funiculi* (columns) on each side of the spinal cord that convey information up and down the cord's tracts. Like the gray matter (which holds the cell bodies of the axons found in the white matter), the columns are named for their location: posterior (or dorsal), anterior (or ventral), and lateral (the section on the sides of the H). The funiculi contain *nerve tracts* that are ascending (transmit sensory information to the brain), descending (transmit motor information from the brain), or bidirectional.

The brain

One of the largest organs in the adult human body, the brain tips the scales at 3 pounds and packs roughly 100 billion neurons (yes, that's billion with a "b") and 900 billion supporting neuroglia cells. In this section, we review six major divisions of the brain from the bottom up: *medulla oblongata, pons, midbrain, cerebellum, diencephalon,* and *cerebrum.* We also discuss the ventricles.

The medulla oblongata

The spinal cord meets the brain at the *medulla oblongata*, or lower half of the *brainstem*. In fact, the medulla oblongata is continuous with the spinal cord at its base (inferiorly). It's located anterior to the cerebellum and extends upwards (superior), posterior to the pons. All the afferent and efferent tracts of the cord can be found in the brainstem as part of two bulges of white matter forming an area referred to as the *pyramids*. Many of the tracts cross from one side to the other at the pyramids, which explains why the right side of the brain controls the voluntary movement of the left side of the body and vice versa.

The medulla oblongata is responsible for maintaining our essential body functions:

>> **Cardiac center:** Controls heart rate

>> **Vasomotor center:** Controls blood pressure

>> **Respiratory center:** Adjusts depth and rate of breathing

The pons

The bulge on the anterior side of the brainstem is the *pons*. The pons is like grand central station, containing (and subsequently routing) numerous different nerve tracts and nuclei (the term for bundles of cell bodies in the brain). As a result, generating a list of functions for the pons would be quite daunting. Some of its more notable jobs, though, include facilitating communication between the cerebrum and the cerebellum, maintaining the regular breathing rate, controlling facial expressions, and causing muscle relaxation during REM sleep (so that we don't act out our dreams).

Along with the pons, the medulla oblongata also forms a network of gray and white matter called the *reticular formation*, found along the most posterior wall of the brainstem. This area is responsible for arousal — making it an essential contributor to the sleep-wake cycle as well as routing sensory information to the higher, or more advanced, parts of the brain. Without the reticular formation, we would never reach (or stay in) consciousness.

The midbrain

Between the pons and the diencephalon lies the *midbrain*, or *mesencephalon*. The midbrain contains the *cerebral aqueduct*, which connects the third and fourth ventricles. (See the section "The ventricles" later in this chapter for more.) It contains the *corpora quadrigemina*, formed by four nuclei. The superior two control visual reflexes while the inferior two control auditory ones. Also in the midbrain is the *red nucleus*, which controls reflexes to maintain posture.

REMEMBER

The medulla oblongata, pons, and midbrain are collectively referred to as the brainstem.

The cerebellum

The *cerebellum* is the second-largest division of the brain. It's just above and overhangs the medulla oblongata and lies just beneath the rear portion of the cerebrum. Inside, the cerebellum resembles a tree called the *arbor vitae*, or "tree of life." The cerebellum is divided into two hemispheres by the *falx cerebelli*, which is an invagination of the dura mater. Numerous nuclei lie within each hemisphere to coordinate the incoming sensory input with the cerebellum's motor output. It receives information from proprioceptors, the eyes, and the semi-circular canals of the ears to assess the body's position and posture. It then sends out motor signals to move the body in the desired manner. It also receives input from the motor cortex of the cerebrum to coordinate our voluntary movements.

The diencephalon

The *diencephalon*, a region between the midbrain and the cerebrum, contains separate brain structures called the *thalamus* and *hypothalamus* as well as the *optic chiasma* (where the optic nerves cross over), the *mammillary bodies* (which play a role in memory recall), and the *pineal gland*.

>> **The thalamus** is an oval structure found just below the *corpus callosum* (the C-shaped band of white matter in the center of the brain that facilitates communication between the right and left halves of the cerebrum). The thalamus is the great integrating center of the brain with the ability to correlate the impulses from tactile, pain, and gustatory (taste) senses with motor reactions.

>> **The epithalamus** contains the *choroid plexus*, a vascular structure that produces cerebrospinal fluid. The pineal gland and olfactory centers also lie within the epithalamus, which forms the roof of the third ventricle.

>> **The hypothalamus** is a cone-shaped structure found between the thalamus and pituitary gland. It contains the control centers for sexual reflexes; body temperature; water, carbohydrate, and fat metabolism; and emotions that affect the heartbeat and blood pressure. Its direct influence on the pituitary gland makes it the bridge between the nervous and endocrine systems.

The diencephalon and the complex set of brain structures surrounding it form the *limbic system*. This is our center of emotions, arousal, and memories. Emotions related to the limbic system include fear, anger, and various emotions related to sexual behavior. Two structures in the limbic system, the amygdala and the hippocampus, play important roles in memories. The amygdala is involved in the storage of memories while the hippocampus sends memories to the appropriate parts of the cerebral hemisphere for long-term storage and retrieves them when necessary.

The cerebrum

The *cerebrum* is mostly white matter surrounded by a thin outer layer of gray matter called the *cerebral cortex*. It features folds or convolutions called *gyri*; furrows and grooves are referred to as *sulci*, and deeper grooves are called *fissures*. A longitudinal fissure separates the cerebral hemispheres into right and left halves. Each hemisphere has a set of controls for sensory and motor activities of the body. Interestingly, it's not just right-side/left-side controls that are reversed in the cerebrum; the upper areas of the cerebral cortex control the lower body activities while the lower areas of the cortex control upper-body activities in a reversal called "little man upside down."

Different functional areas of the cerebral cortex are divided into lobes which are named for the skull bones they lie beneath. They are further divided into sensory, motor, and association areas (that provide memory and reasoning).

>> **Frontal lobe:** The most advanced part of the brain. It is responsible for our highest intellectual functions such as planning, logical reasoning, morality, and social behaviors. It contains the *primary motor cortex* which initiates voluntary movements. The frontal lobe also contains *Broca's area* which is responsible for the generation of speech.

>> **Parietal lobe:** The primary site for integration of sensory information. It is the area responsible for our conscious perception of touch, temperature, and pain. The *somatosensory cortex* is located here, as is *Wernicke's area,* which is responsible for language comprehension. The parietal lobe is also responsible for spatial awareness and maintaining attention.

>> **Temporal lobe:** The primary auditory center. It houses situational memories such as visual scenes and music. It also plays a role in understanding and producing speech.

>> **Occipital lobe:** The primary vision center. It processes all visual information and provides meaning to it by connecting with memories.

The ventricles

The meninges of the brain provide tight control over what molecules from the bloodstream are allowed into the tissues of the brain and spinal cord. As a result, the CNS circulates CSF (cerebrospinal fluid) to provide nutrients and remove wastes. Structures within the brain, the *ventricles,* allow for this. The largest of the four, the *right and left lateral ventricles,* lie within the cerebrum and create most of the CSF. The *third ventricle* is found within the diencephalon and receives CSF from the lateral ventricles. It then sends the CSF to the *fourth ventricle* via the *cerebral aqueduct.* The fourth ventricle is in the brainstem and is continuous with the central canal of the spinal cord. It also passes CSF to the subarachnoid space.

Cranial nerves

Twelve pairs of cranial nerves connect to the central nervous system via the brain (as opposed to the 31 pairs that connect via the spinal cord). Cranial nerves are identified by Roman numerals I through XII, and memorizing them is a classic test of anatomical knowledge. Check out Table 9-2 for a listing of all the nerves.

Table 9-2   Cranial Nerves

Number	Name	Type	Function
I	Olfactory	Sensory	Smell
II	Optic	Sensory	Vision
III	Oculomotor	Mixed nerve	Eyeball muscles
IV	Trochlear	Mixed nerve	Eyeball muscles
V	Trigeminal	Tri means "three," so the three types of trigeminal nerves are 1) Opthalmic nerve: sensory nerve; skin and mucous membranes of face and head; 2) Maxillary nerve: mixed nerve; mastication; 3) Mandibular nerve: mixed nerve; mastication	Skin; mastication (chewing)

Number	Name	Type		Function
VI	Abducens	Mixed nerve		Eye movements laterally
VII	Facial	Mixed nerve		Facial expression; salivary secretion; taste
VIII	Vestibulocochlear	Sensory		Auditory nerve for hearing and equilibrium
IX	Glossopharyngeal	Mixed nerve		Taste; swallowing muscles of pharynx; salivation; gag reflex; speech
X	Vagus	Mixed nerve		Controls most internal organs (viscera) from head and neck to transverse colon
XI	Accessory	Mixed nerve		Swallowing and phonation; rotation of head and shoulder
XII	Hypoglossal	Motor nerve		Tongue movements

TIP

The first letters of each of these nerve names, in order, are OOOTTAFVGVAH. That's a mouthful, but students have come up with a number of memory tools to remember them. Our favorite is Old Opera Organs Trill Terrific Arias For Various Grand Victories About History.

Put your knowledge of the central nervous system to the test:

EXAMPLE

Q. The meninges' functions are primarily

    a. immunological.

    b. supportive.

    c. protective.

    d. a and b.

    e. b and c.

A. The correct answer is b and c (supportive and protective). Yes, meninges have two functions.

---

 30 The cerebrum is divided into two major halves, called _____, by the _____.

    a. cerebellar hemispheres, transverse fissure

    b. cerebral spheres, central sulcus

    c. cerebellar spheres, lateral sulcus

    d. cerebral hemispheres, longitudinal fissure

 31 You've been in an accident, and you have injured the left side of your cerebrum. Which areas of your body would you most likely have trouble moving?

    a. The right side

    b. The lower body

    c. The upper extremities

    d. The left side

32 Where within the cerebrum does visual activity take place?

    a. Parietal lobe

    b. Temporal lobe

    c. Occipital lobe

    d. Frontal lobe

33 What, among many things, does the cerebellum do?

    a. Controls visual activity

    b. Generates associative reasoning

    c. Interprets auditory stimulation

    d. Coordinates motor control

34-38 Match the term to its description.

    a. Has the capacity to arouse the brain to wakefulness

    b. Myelinated fibers

    c. Coordinates communication between right and left brain

    d. Bundle of neuron cell bodies

    e. Unmyelinated fibers, neuron cell bodies, and neuroglia

34 _____ White matter

35 _____ Reticular formation

36 _____ Ganglion

37 _____ Corpus callosum

38 _____ Gray matter

39-43 Match the structure to its description.

    a. Bridge connecting the medulla oblongata and cerebellum

    b. Contains the centers that control cardiac, respiratory, and vasomotor functions

    c. Controls visual and auditory reflexes

    d. Controls motor coordination and refinement of muscular movement

    e. Controls sensory and motor activity of the body

39 _____ Pons

40 _____ Cerebellum

41 _____ Medulla oblongata

42 _____ Cerebrum

43 _____ Midbrain

Taking Side Streets: The Peripheral Nervous System

44 What does the choroid plexus do?

    a. Filters cerebrospinal fluid

    b. Absorbs cerebrospinal fluid

    c. Produces cerebrospinal fluid

    d. Eliminates exhausted cerebrospinal fluid

45-57 Use the terms that follow to identify the parts of the brain shown in Figure 9-3.

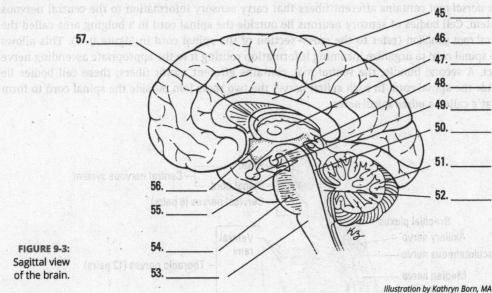

45. _____

46. _____

47. _____

48. _____

49. _____

50. _____

51. _____

52. _____

57. _____

56. _____

55. _____

54. _____

53. _____

FIGURE 9-3:
Sagittal view
of the brain.

Illustration by Kathryn Born, MA

    a. Pons

    b. Thalamus

    c. Cerebellum

    d. Corpus callosum

    e. Third ventricle

    f. Hypothalamus

    g. Cerebrum

    h. Cerebral aqueduct

    i. Midbrain

    j. Pituitary gland

    k. Medulla oblongata

    l. Fourth ventricle

    m. Mammillary body

Taking Side Streets: The Peripheral Nervous System

The *peripheral nervous system* (PNS) is the network that carries information to and from the spinal cord. Among its key structures are 31 pairs of spinal nerves (see Figure 9-4). Also included in the PNS are all the individual neuron receptors that relay information as well as some glial cells. Eight of the spinal nerve pairs are *cervical* (having to do with the neck), 12 are *thoracic* (relating to the chest, or thorax), five are *lumbar* (between the lowest ribs and the pelvis), five are *sacral* (the posterior section of the pelvis), and one is *coccygeal* (relating to the tailbone). Spinal nerves connect with the spinal cord by two bundles of nerve fibers, or *roots*. The *dorsal root* contains afferent fibers that carry sensory information to the central nervous system. Cell bodies of sensory neurons lie outside the spinal cord in a bulging area called the *dorsal root ganglion* (refer to the cross-section of the spinal cord in Figure 9-2). This allows the spinal cord to organize incoming information routing it to the appropriate ascending nerve tract. A second bundle, the *ventral root*, contains efferent motor fibers; these cell bodies lie inside the spinal cord. In each spinal nerve, the two roots join outside the spinal cord to form what's called a *mixed spinal nerve*.

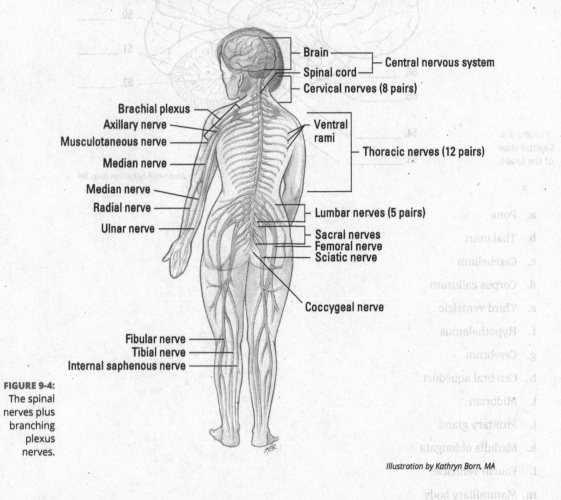

FIGURE 9-4: The spinal nerves plus branching plexus nerves.

Brain
Spinal cord
Central nervous system
Cervical nerves (8 pairs)
Brachial plexus
Axillary nerve
Musculotaneous nerve
Ventral rami
Median nerve
Thoracic nerves (12 pairs)
Median nerve
Radial nerve
Ulnar nerve
Lumbar nerves (5 pairs)
Sacral nerves
Femoral nerve
Sciatic nerve
Coccygeal nerve
Fibular nerve
Tibial nerve
Internal saphenous nerve

Illustration by Kathryn Born, MA

Spinal reflexes, or *reflex arcs*, occur when a sensory neuron transmits a "danger" signal — like a sensation of burning heat — through the dorsal root ganglion. An interneuron (or association neuron) in the spinal cord receives the stimulus and passes along the signal directly to a motor neuron (or efferent fiber) that stimulates a muscle, which immediately pulls the burning body part away from heat (see Figure 9-5). Other reflexes, like the knee-jerk, or *patellar reflex*, forego the interneuron and the sensory one directly stimulates the motor one.

After a spinal nerve leaves the spinal column, it divides into two small branches. The posterior, or *dorsal ramus*, goes along the back of the body to supply a specific segment of the skin, bones, joints, and longitudinal muscles of the back. The ventral, or *anterior ramus*, is larger than the dorsal ramus and supplies the anterior and lateral regions of the trunk and limbs.

Groups of spinal nerves interconnect to form an extensive network called a *plexus* (Latin for "braid"), each of which connects through the anterior ramus, including the *cervical plexus* of the neck, *brachial plexus* of the shoulder and axilla, and *lumbosacral plexus* of the lower back (including the body's largest nerve, the *sciatic nerve*). However, there's no plexus in the thoracic area. Instead, the anterior ramus directly supplies the intercostal muscles (literally "between the ribs") and the skin of the region.

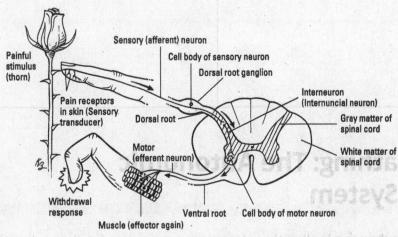

FIGURE 9-5:
A reflex arc —
responding
to pain.

Illustration by Kathryn Born, MA

---

58 What enters the nervous system through a dorsal root ganglion?

   a. A dorsal ramus

   b. An anterior ramus

   c. An efferent nerve

   d. An afferent nerve

   e. The lumbosacral plexus

59 What is a mixed spinal nerve?

  a. The point outside the spinal cord where the dorsal root and the ventral root join

  b. The bulging area inside the spinal cord where afferent fibers deliver sensory information

  c. A neuromere that interacts with a plexus

  d. The point along the spinal cord where the dorsal ramus and the anterior ramus join

  e. A nerve that runs in and then directly out of the spinal cord creating a reflex arc

60 Networks of nerves often form a plexus, from the Latin word for "braid." Which spinal nerve area does not?

  a. Cervical

  b. Lumbar

  c. Sacral

  d. Thoracic

Keep Breathing: The Autonomic Nervous System

Just as the name implies, the autonomic nervous system functions automatically. Divided into the *sympathetic* and *parasympathetic* systems, it activates the involuntary smooth and cardiac muscles and glands to carry out all the body functions that happen automatically. (Autonomic, automatic — see the connection?) Autonomic functions are under the control of the hypothalamus, cerebral cortex, brainstem.

The sympathetic system, which is responsible for the body's involuntary fight-or-flight response to stress, is defined by the autonomic fibers that exit the thoracic and lumbar segments of the spinal cord. The parasympathetic system (affectionately referred to as "rest and digest") is defined by the autonomic fibers that either exit the brainstem via the cranial nerves or exit the sacral segments of the spinal cord. (You can see both systems in Figure 9-6.)

REMEMBER

The sympathetic and parasympathetic systems oppose each other in function, helping to maintain *homeostasis*, or balanced activity in the body systems. Yet, often the sympathetic and parasympathetic systems work in concert. The sympathetic system dilates the eye's pupil, but the parasympathetic system contracts it again. The sympathetic system quickens and strengthens the heart while the parasympathetic slows the heart's action. The sympathetic system contracts blood vessels in the skin so more blood goes to muscles for a fight-or-flight reaction to stress, and the parasympathetic system dilates the blood vessels when the stress concludes.

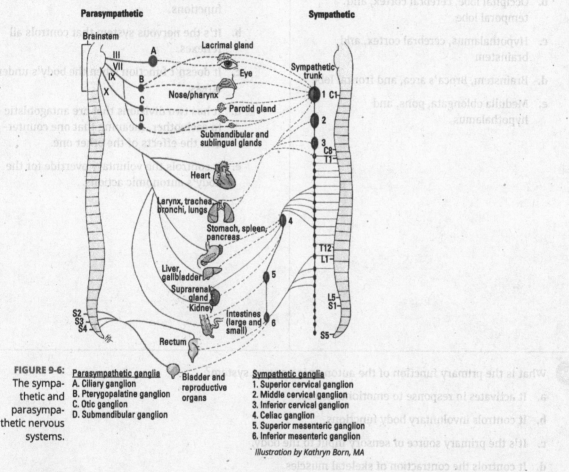

FIGURE 9-6: The sympathetic and parasympathetic nervous systems.

Parasympathetic ganglia
A. Ciliary ganglion
B. Pterygopalatine ganglion
C. Otic ganglion
D. Submandibular ganglion

Sympathetic ganglia
1. Superior cervical ganglion
2. Middle cervical ganglion
3. Inferior cervical ganglion
4. Celiac ganglion
5. Superior mesenteric ganglion
6. Inferior mesenteric ganglion

Illustration by Kathryn Born, MA

REMEMBER

The motor division of the nervous system (all outgoing information) is split into the autonomic and *somatic nervous system*. This carries out the voluntary orders for skeletal muscles to contract.

See whether any of the following practice questions touch a nerve:

61 Which three parts of the brain control autonomic functions?

a. Medulla oblongata, parietal lobe, and cerebellum

b. Occipital lobe, cerebral cortex, and temporal lobe

c. Hypothalamus, cerebral cortex, and brainstem

d. Brainstem, Broca's area, and frontal lobe

e. Medulla oblongata, pons, and hypothalamus

62 Which of the following statements is true about the autonomic nervous system?

a. It has two parts: the parasympathetic that carries out the stress response, and the sympathetic that controls all normal functions.

b. It's the nervous system that controls all reflexes.

c. It doesn't function when the body's under stress.

d. It has two divisions that are antagonistic to each other, meaning that one counteracts the effects of the other one.

e. It controls the voluntary override for the body's autonomic actions.

63 What is the primary function of the autonomic nervous system?

a. It activates in response to emotional stress.

b. It controls involuntary body functions.

c. It's the primary source of sensory input to the body.

d. It controls the contraction of skeletal muscles.

e. It works to block signals that may overload the overall nervous system.

Coming to Your Senses

The nervous system must have some way to perceive its environment in order to generate appropriate responses. That's where the senses come in. *Sense receptors* are the specialized nervous cells that respond to stimuli — like increased temperature, bitter tastes, and sharp points — by generating a nerve impulse. Although there are millions of *general sense receptors* found throughout the body that can convey touch, pain, and physical contact, there are far fewer of the *special sense receptors* — those located in the head — that really bring meaning to your world. Here we discuss vision, hearing, and equilibrium. The special senses of olfaction (smell) and gustation (taste) are discussed in Chapters 13 and 14, respectively.

REMEMBER

Sense receptors are classified both by the location of the stimulus as well as their mechanism of action. With respect to location, these cells can be *exteroreceptors*, which respond to stimulus on the body surface; *interoreceptors*, which monitor internal stimuli; and *telereceptors*, which respond to stimuli without physical contact (sight, hearing, smell). Table 9-3 shows the major categories of receptors based on the type of stimulus they respond to.

Table 9-3   Sensory Receptors

Receptor Type	Stimulus	Example
Chemoreceptor	Change in chemical concentration	Olfactory, pH
Nociceptor	Tissue damage	pain
Thermoreceptor	Temperature change	Hot and cold
Mechanoreceptor	Pressure/mechanical forces	Touch, hearing
Photoreceptor	Light	Rods and cones
Proprioceptor	Stretch	Muscle spindles

On watch: The eyes

Although there are many romantic notions about eyes, the truth is that an eyeball is simply a hollow sphere bounded by a trilayer wall and filled with a gelatinous fluid called, oddly enough, *vitreous humor*. The outer fibrous coat is made up of the *sclera* (the white of the eye) and the *cornea* in front. The sclera provides mechanical support, protection, and a base for attachment of eye muscles, which assist in the focusing process. The cornea covers the anterior with a clear window.

An intermediate, or vascular, layer called the *uvea* provides blood and lymphatic fluids to the eye; regulates the amount of light entering the eye; and secretes and reabsorbs *aqueous humor*, a thin, watery liquid that fills the anterior chamber of the eyeball in front of the iris and the posterior chamber between the iris and the lens. The uvea has three components:

>> **The iris:** Contains blood vessels and smooth muscle fibers to control the pupil's diameter. The pigments that provide our eye color prevent light from entering anywhere other than the pupil.

>> **The choroid coat:** A thin, dark brown, vascular layer lining most of the sclera on the back and sides of the eye. The choroid contains arteries, veins, and capillaries that supply the *retina* and sclera with nutrients, and it also contains pigment cells to absorb light and prevent reflection and blurring.

>> **The ciliary body:** An extension of the choroid coat that leads to the iris. It also attaches to the lens via *suspensory ligaments* and controls the lens' shape to aid in focusing.

The crystalline lens consists of concentric layers of protein. It's *biconvex* in shape, bulging outward and with help from the cornea, bends light projecting it into the *inner tunic*. When the muscles of the ciliary body contract, the shape of the lens changes, altering the visual focus. This process of *accommodation* allows the eye to see objects both at a distance and close-up.

The inner tunic is the internal lining of the eye. Here, the *retina* lies atop the choroid coat containing the photoreceptors that will communicate with the occipital lobe of the brain. The space created, called the *posterior cavity,* is filled with *vitreous humor*, a jelly-like fluid that maintains the shape of the eye. The photoreceptors of the retina come in two varieties: *rods* and *cones*, whose axons are bundled into the optic nerve that exits out the back. The rods are dim light receptors that provides the general shape of the image. The cones detect bright light and provide sharpness and color to the image. The retina has an *optic disc*, which is essentially a blind spot incapable of producing an image. The eye's blind spot is a result of the absence of photoreceptors in the area of the retina where the optic nerve leaves the eye.

The macula or *macula lutea* (Latin for "yellow spot") is an oval-shaped, highly pigmented yellow area near the center of the retina. There is a small dimple near the center of the macula called the *fovea centralis* (*fovea* is Latin for "pit"). This is where the eye's vision is sharpest and it is where most of the eye's color perception occurs. Only cones can be found in this area, smaller and more closely packed than anywhere else in the retina. The fovea sees only the central 2 degrees of the visual field, which encompasses roughly twice the width of your thumbnail at arm's length. If an observed area or object is larger than this angle of vision, such as when reading, the eyes must constantly shift their gaze, a process known as *saccadic movement,* to bring different portions of the image into the fovea. For this reason, the cognitive brain is responsible for oculomotor commands and piecing together a comprehensive visual image.

The *palpebrae* (eyelids) extend from the edges of the eye orbit, into which roughly five-sixths of the eyeball is recessed. A mucous membrane called the *conjunctiva* covers the inner surface of each eyelid and the anterior surface of the eye, but not the cornea. (The infamous "pink eye" is the inflammation of this membrane.) Up top and to the side of the orbital cavities are lacrimal glands that secrete tears that are carried through a series of lacrimal ducts to the conjunctiva of the upper eyelid. Ultimately, secretions drain from the eyes through the nasolacrimal ducts.

Listen up: The ears

Human ears — otherwise called *vestibulocochlear* organs — are more than just organs of hearing. They also serve as organs of equilibrium, or balance. Here are the three divisions of the ear:

>> **The external ear** includes the *auricle,* or *pinna,* which is the recognizable folded appendage made of elastic cartilage and skin. Extending into the skull is the *ear canal,* or *external auditory meatus,* a short passage through the temporal bone ending at the *tympanic membrane,*

or *eardrum*. Sebaceous glands near the external opening and *ceruminous* glands in the upper wall produce the brownish substance known as earwax, or *cerumen*.

» **The middle ear** is a small, usually air-filled cavity in the skull that's lined with mucous membrane. It contains the *Eustachian tube* that connects with the pharynx to keep air pressure equal on both sides of the tympanic membrane. Three small bones called *auditory ossicles* occupy the middle ear, deriving their names from their shapes: the *malleus* (hammer), the *incus* (anvil), and the *stapes* (stirrup).

» **The internal ear** has three main components: the *semi-circular canals*, the *cochlea*, and the *vestibule*, which is the space between them. The cochlea is the snail-shaped structure that contains the *hair cells* — the mechanoreceptors that transmit auditory input to the temporal lobe. The semi-circular canals are three fluid-filled tubes that communicate information about the body's position to the cerebellum via the vestibulocochlear nerve. When we move, tilting to one side for example, the fluid inside does not. This stimulates the proprioceptors lining the canals.

The process of hearing a sound follows these basic steps:

REMEMBER 1. Sound waves travel through the auditory canal, striking the eardrum, making it vibrate.

2. This vibration is passed through the malleus, the incus, and finally the stapes.

3. The stapes then vibrates the oval window of the vestibular canal, translating the motion into the *perilymph fluid* of the cochlea.

4. The vibrating fluid begins stimulating the *endolymph fluid* in the membranous area of the cochlea.

5. The endolymph fluid in turn stimulates the hair cells, which transmit the impulses to the brain over the auditory (cochlear) nerve.

EXAMPLE

Q. The most sensitive region of the retina producing the greatest visual acuity is the

a. optic disc.

b. cornea.

c. fovea centralis.

d. choroid coat

e. macula lutea.

A. The correct answer is fovea centralis. It's loaded with light-sensitive cones.

Use the terms that follow to identify the internal structures of the eye shown in Figure 9-7.

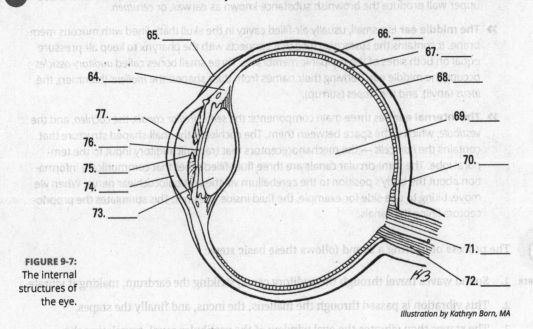

65. _____   66. _____
67. _____
64. _____   68. _____
77. _____   69. _____
76. _____
75. _____   70. _____
74. _____
73. _____

FIGURE 9-7:
The internal
structures of
the eye.

71. _____
72. _____

Illustration by Kathryn Born, MA

a. Optic disc

b. Pupil

c. Optic nerve

d. Retina

e. Cornea

f. Sclera

g. Ciliary body

h. Fovea centralis

i. Lens

j. Anterior chamber (contains aqueous humor)

k. Choroid coat

l. Conjunctiva

m. Iris

n. Posterior cavity (contains vitreous humor)

78 Outline the path of light through the eye.

_____

_____

_____

_____

79 The accommodation, or focusing, of the eye involves which of these?

a. Sphincter of the pupil

b. Contraction of the iris

c. Action of the ciliary muscles

d. Adjustment of the cornea

e. Contraction of the pupil

80 All the ganglion cells of the retina merge and exit the eye forming the optic nerve. What is the area where they exit called?

a. Macula lutea

b. Optic disc

c. Choroid coat

d. Vitreous humor

e. Fovea centralis

81-88 Sound waves travel through the ear canal and cause the **81._____**, or eardrum, to vibrate. Those vibrations then are picked up by three small bones, called **82._____**. The **83._____** is the first to receive the vibrations. It then passes them to the **84._____** and then the **85._____**. These vibrations are then transferred to the **86._____**, where fluid is jostled stimulating the **87._____**. These communicate the sounds to the auditory cortex in the **88._____** of the brain.

a. Acoustic meatus

b. Cochlea

c. Hair cells

d. Incus

e. Malleus

f. Occipital lobe

g. Ossicles

h. Semicircular canals

i. Stapes

j. Temporal lobe

k. Tympanic membrane

l. Vestibule

89 What is the difference between aqueous humor and vitreous humor?

    a. One lines the sclera, and the other bathes the iris.

    b. One is thin and watery, and the other is thick and gelatinous.

    c. One acts as a photoreceptor, and the other acts as a focusing mechanism.

    d. One is subtly funny, and the other is brashly obvious.

90 What is the role of the semicircular canals?

    a. provide structure and support to the orbit

    b. contain rods, cones, and hair cells to receive stimulus

    c. translate vibration of the ossicles to sound

    d. equalize pressure in the ear's chambers

    e. transmit information on the body's position

Answers to Questions on the Nervous System

The following are answers to the practice questions presented in this chapter.

(1-5) Fill in the blanks. The communicating cells of the nervous system, or **1. neurons** have two essential properties. First is **2. irritability**, the tendency to respond to a stimulus. Second is the ability to generate electricity, called **3. conductivity**. This flow of electricity is known as **4. impulse** but cannot be used to communicate with other neurons. For this, **5. neurotransmitters** are used.

(6) What are the three primary functions of the nervous system? **c. Perception, integration, action.** Each of the other answer options includes at least one function that clearly belongs to another of the body's systems.

(7) Which structure is found at the end of an axon? **e. synaptic knob.** Though you could argue that soma is true as well, but that's not really the end of the axon, that's the beginning.

(8) Which part of a neuron is responsible for conducting electrical signals toward the cell body? **d. dendrite**

(9) Which membrane surrounds an axon (nerve fiber)? **b. Neurilemma.** It actually wraps around the myelin sheath, so it's always on the outside.

(10) Astrocytes: **f. Cells that contribute to the repair process of the central nervous system.**

(11) Microglia: **d. Cells that are phagocytic.**

(12) Oligodendrocytes: **b. Cells that form and preserve myelin sheaths in CNS.**

(13) Schwann cell: **a. Cells that form and preserve myelin sheaths in PNS.**

(14) Satellite cell: **c. Cells that provide nutrients to bundled cell bodies.**

(15) Ependyma: **e. Cells that form linings in the CNS.**

(16) How many dendrites and axons are there on a multipolar neuron? **a. Many dendrites and one axon.** A neuron may have one, many, or no dendrites, but it always has a single axon.

(17) A synapse between neurons is best described as the transmission of **d. an impulse through a chemical change.** With all that acetylcholine and cholinesterase floating around, it must be a chemical transmission.

(18) The connective tissue that surrounds a bundle of fibers in a nerve is known as the **b. perineurium.**

(19) Nervous tissue is essentially neurons connected end-to-end with very little matrix. **False.** Abundant space is visible between neurons, dotted with glial cells and numerous "fibers" which are actually axons of other neurons.

(20) Which statement best describes how neurons communicate? **d. In response to stimulus, the dendrites generate an electrical signal which is passed to the cell body. The axon will then spread the impulse down its length releasing a chemical at the end.** Note that the electrical signal in a neuron is not termed impulse until it is in the axon.

(21) All-or-none response: **c. Threshold of excitation determines ability to respond**

(22) Hyperpolarization: **d. A drop in charge inside the cell past resting potential**

(23) Anion: **b. Negatively charged molecule concentrated inside the cell**

(24) Polarization: **a. A differential in charge from outside of the cell to the inside**

(25) Depolarization: **e. An increase in charge inside of the cell**

(26) Place the steps of action potential in order: **8. Potassium gates close, 10. Pumps restore polarization, 3. Cell depolarizes, 2. Sodium rushes in, 6. Potassium flows out, 1. Sodium gates open, 4. Sodium gates close, 5. Potassium gates open, 11. Ions return to initial concentrations, 7. Cell repolarizes, 9. Cell is hyperpolarized. Though steps 4 and 5 happen simultaneously (Na+ gates open, K+ gates close), it makes more sense to address the open sodium gates first.**

(27) What is the role of Ca2+ ions in nervous communication? **e. Enters the cell to trigger the release of neurotransmitters**

(28) Neurotransmitters bind to receptors on the dendrite opening channels for positive ions to rush in and depolarize the cell: **False. This is a bit of a trick question because that's exactly what excitatory neurotransmitters do. But there are also inhibitory neurotransmitters which can lead to hyperpolarization instead.**

(29) Explain the difference between impulse and action potential. **Impulse is the step-wise process that occurs in a single segment of axon. It begins with voltage-gated sodium channels opening, leading to local depolarization, and ends with the sodium/potassium pumps returning the ions to their original positions. However, as the sodium ions rush in and depolarize that segment of axon, the threshold is reached in the next segment, opening the next set of sodium channels. So the steps early in each action potential triggers the start of the next one. This progression of action potentials constitutes the impulse.**

(30) The cerebrum is divided into two major halves, called **d. cerebral hemispheres,** by the **d. longitudinal fissure.** Cerebrum = cerebral, and two halves = hemispheres. You can remember the fissure's name by equating it to Earth's prime meridian, which separates the Eastern and Western Hemispheres, and longitudinal is the most likely position for an equal division.

(31) You've been in an accident, and you have injured the left side of your cerebrum. Which areas of your body would you most likely have trouble moving? **a. The right side.** In the cerebrum, right = left and up = down. Clear as mud?

(32) Where within the cerebrum does visual activity take place? **c. Occipital lobe.** To remember, use the word "occipital" to bring to mind the word "optic," which of course is related to visual activity.

(33) What, among many things, does the cerebellum do? **d. Coordinates motor control.** As the second-largest division of the brain, the cerebellum also is known as the "little brain."

(34) White matter: **b. Myelinated fibers.**

(35) Reticular formation: **a. Has the capacity to arouse the brain to wakefulness.**

(36) Ganglion: **d. Bundle of neuron cell bodies.**

(37) Corpus Callosum: **c. Coordinates communication between right and left brain.**

(38) Gray matter: **e. Unmyelinated fibers, neuron cell bodies, and neuroglia.**

(39) Pons: **a. Bridge connecting the medulla oblongata and cerebellum.**

(40) Cerebellum: **d. Controls motor coordination and refinement of muscular movement.**

(41) Medulla oblongata: **b. Contains the centers that control cardiac, respiratory, and vasomotor functions.**

(42) Cerebrum: **e. Controls sensory and motor activity of the body.**

(43) Midbrain: **c. Controls visual and auditory reflexes.**

(44) What does the choroid plexus do? **c. Produces cerebrospinal fluid.** This one requires rote memorization — sorry!

(45-57) Following is how Figure 9-3, the brain, should be labeled:

45. **g. Cerebrum;** 46. **d. Corpus callosum;** 47. **b. Thalamus;** 48. **m. Mammillary body;** 49. **j. Pituitary gland;** 50. **h. Cerebral aqueduct;** 51. **c. Cerebellum;** 52. **l. Fourth ventricle;** 53. **k. Medulla oblongata;** 54. **a. Pons;** 55. **i. Midbrain;** 56. **f. Hypothalamus;** 57. **e. Third ventricle.**

(58) What enters the nervous system through a dorsal root ganglion? **d. an afferent nerve.** The dorsal root brings in sensory information. Besides, all the other terms are efferent structures.

(59) What is a mixed spinal nerve? **a. The point outside the spinal cord where the dorsal root and the ventral root join.** That juncture occurs in each of the 31 pairs of spinal nerves.

(60) Networks of nerves often form a plexus, from the Latin word for "braid." Which spinal nerve area does not? **d. Thoracic.** There is far less nerve networking to be "braided" there.

(61) Which three parts of the brain control autonomic functions? **c. Hypothalamus, cerebral cortex, and brainstem.** We may have thrown you for a loop by calling the medulla oblongata and the pons by their more common name: brainstem. Beware of Choice e: While associated with advanced functions, the cerebral cortex plays a role here too.

(62) Which of the following statements is true about the autonomic nervous system? **d. It has two divisions that are antagonistic to each other, meaning that one counteracts the effects of the other one.** As a result, the body achieves homeostasis.

(63) What is the primary function of the autonomic nervous system? **b. It controls involuntary body functions.** That's the only answer option with a sense of automation.

(64-77) Following is how Figure 9-7, the internal structures of the eye, should be labeled.

64. **g. Ciliary body;** 65. **l. conjunctiva;** 66. **f. Sclera;** 67. **d. Retina;** 68. **k. Choroid coat;** 69. **n. Posterior cavity (filled with vitreous humor);** 70. **h. Fovea centralis;** 71. **a. Optic disc;** 72. **c. Optic nerve;** 73. **i. Lens;** 74. **e. Cornea;** 75. **m. Iris;** 76. **b. Pupil;** 77. **j. Anterior chamber (contains aqueous humor).**

(78) Outline the path of light through the eye. **Cornea, anterior chamber, pupil, posterior chamber, lens, posterior cavity, retina.**

You may want to develop your own mnemonic for this path. Here's an unwieldy example: Continuous Air Conditioning Puts Possible Charges Lower on the Politically Correct Radar.

TIP

79 The accommodation, or focusing, of the eye involves which of these? **c. Action of the ciliary muscles.** They reshape the lens by contracting and relaxing as needed to bring things into focus.

80 All the ganglion cells of the retina merge and exit the eye forming the optic nerve. What is the area where they exit called? **b. Optic disc.** Sometimes referred to as the *optic nerve head*, the optic disc has no photoreceptors of its own, causing a "blind spot" in each of your eyes.

81–88 Sound waves travel through the ear canal and cause the **k. tympanic membrane**, or eardrum, to vibrate. Those vibrations then are picked up by three small bones, called **g. ossicles.** The **e. malleus** is the first to receive the vibrations. It then passes them to the **d. incus** and then the **i. stapes.** These vibrations are transferred to the **b. cochlea**, where fluid is jostled stimulating the **c. hair cells.** These communicate the sounds to the auditory cortex in the **j. temporal lobe** of the brain.

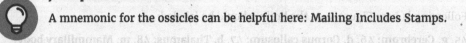

TIP A mnemonic for the ossicles can be helpful here: Mailing Includes Stamps.

89 What is the difference between aqueous humor and vitreous humor? **b. One is thin and watery, and the other is thick and gelatinous.** Vitreous humor (which is thick and gelatinous) fills the hollow sphere of the eyeball, but aqueous humor (which is thin and watery) fills two chambers toward the front of the eyeball.

90 What is the role of the semicircular canals? **e. transmit information on the body's position**

Chapter **10**

Raging Hormones: The Endocrine System

The human body has two separate command and control systems that work in harmony most of the time but also work in very different ways. Designed for instant response, the nervous system cracks its cellular whip using electrical signals that make entire systems hop to their tasks with no delay (you can get a feel for the nervous system in Chapter 9). By contrast, the endocrine system's glands use chemical signals called *hormones* that behave like the steering mechanism on a large, fully loaded ocean tanker; small changes can have big impacts, but it takes quite a bit of time for any evidence of the change to make itself known. At times, parts of the nervous system stimulate or inhibit the secretion of hormones, and some hormones are capable of stimulating or inhibiting the flow of nerve impulses.

The word "hormone" originates from the Greek word *hormao*, which literally translates as "I excite." And that's exactly what hormones do. Each chemical signal stimulates specific parts of the body, known as *target tissues* or *target cells*. The body needs a constant supply of hormonal signals to grow, maintain homeostasis, reproduce, and conduct myriad processes.

In this chapter, we go over which glands do what and where, as well as review the types of chemical signals that play various roles in the body. You also get to practice discerning what the endocrine system does, how it does it, and why the body responds like it does.

No Bland Glands

REMEMBER

Technically, there are ten or so primary endocrine glands, made up of epithelial cells embedded within connective tissue. There are also various other hormone-secreting tissues scattered throughout the body. *Exocrine glands*, like the sweat and mammary glands, release their secretions though ducts destined for a body surface. (Keep in mind your body has both external and internal surfaces.) Like endocrine glands, the *paracrine glands* secrete their product into the interstitial fluid. The *paracrine factors* then bind to receptors on nearby cells inducing their response. Secretions from the endocrine glands, though, are pulled into capillaries for destinations elsewhere in the body. So our hormones make the journey to their target cells via the bloodstream without errantly triggering an unrelated event along the way. This is due to their structural specificity; that is, each different hormone has a unique shape that seems tailor-made for the receptors of its target cells.

Hormones can be classified either as *steroid* (derived from cholesterol) or *nonsteroid* (derived from amino acids and other proteins). The steroid hormones — which include testosterone, estrogen, and cortisol — are the ones most closely associated with emotional outbursts and mood swings. Because they are lipid-soluble, steroid hormones can enter just about any cell in the body. However, to trigger the desired response, a receptor molecule inside of the cell must bind the hormone and carry it to the nucleus.

Nonsteroid hormones are divided among four classifications:

>> Some are *modified amino acids,* including such things as epinephrine and norepinephrine, as well as melatonin.

>> Others are *peptide*-based (at least three amino acids), including antidiuretic hormone (ADH) and oxytocin (OT).

>> *Glycoprotein*-based hormones include follicle-stimulating hormone (FSH), and luteinizing hormone (LH) — both closely associated with the reproductive systems (which we cover in Chapters 16 and 17).

>> *Protein*-based nonsteroid hormones include such crucial substances as insulin and growth hormone (GH) as well as prolactin and parathyroid hormone.

Since these hormones are water-soluble, they are unable to enter the target cell. Instead, they bind to receptors on the cell's surface triggering a cascade of chemical reactions using a *second messenger* inside of the cell, ultimately leading to the desired effect.

REMEMBER

Hormone functions include controlling the body's internal environment by regulating its chemical composition and volume, activating responses to changes in environmental conditions to help the body cope, influencing growth and development, enabling several key steps in reproduction, regulating components of the immune system, and regulating organic metabolism.

See if all this hormone-speak is sinking in:

1-5 **Mark the statement with a T if it's true or an F if it's false:**

1 _____ The endocrine system brings about changes in the metabolic activities of the body tissue.

2 _____ The amount of hormone released is determined by the body's need for that hormone at the time.

3 _____ The glands of the endocrine system are composed of cartilage cells.

4 _____ Endocrine glands aren't functional in reproductive processes.

5 _____ Some hormones can be derivatives of amino acids, whereas others are synthesized from cholesterol.

6 _____ glands secrete their product through ducts while _____ glands secrete their product into the interstitial fluid, which flows into the blood.

a. Exocrine; endocrine

b. Endocrine; exocrine

c. Endocrine; paracrine

d. Paracrine, heterocrine

e. Exocrine; paracrine

7 **Explain how, due to their structural differences, steroid and nonsteroid hormones elicit responses from their target cells in different ways.**

_____

_____

_____

_____

Enter the Ringmasters

The key glands of the endocrine system include the *pituitary* (also called the *hypophysis*), *adrenal* (also referred to as *suprarenal*), *thyroid, parathyroid, thymus, pineal, islets of Langerhans* (within the *pancreas*), and *gonads* (testes in the male and ovaries in the female). But of all these, it's the pituitary working in concert with the hypothalamus in the brain that really keeps things rolling (see Figure 10-1).

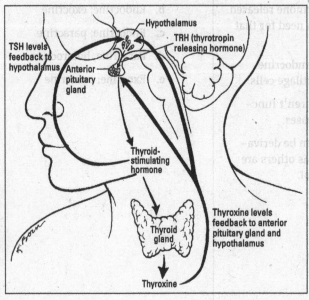

FIGURE 10-1: The working relationship of the hypothalamus and the pituitary gland.

Illustration by Kathryn Born, MA

The hypothalamus

REMEMBER

The hypothalamus is the unsung hero linking the body's two primary control systems — the endocrine system and the nervous system. Part of the brain and part of the endocrine system, the hypothalamus is connected to the pituitary via a narrow stalk called the *infundibulum*. In its supervisory role, the hypothalamus provides neurohormones to control the pituitary gland's secretions. Releasing and inhibiting hormones produced by the hypothalamus regulate the release of most anterior pituitary hormones.

The hypothalamus sits just above the pituitary gland, which is nestled in the middle of the human head in a depression of the skull's sphenoid bone called the *sella turcica.*

The pituitary

The pituitary's *anterior lobe*, also called the *adenohypophysis*, is sometimes called the "master gland" because of its role in regulating and maintaining other endocrine glands. Hormones that act on other endocrine glands are called *tropic hormones*; all the hormones produced in the anterior lobe are polypeptides. Among the hormones produced in the anterior lobe of the pituitary gland are the following:

>> **Follicle-stimulating hormone (FSH):** Signals ovum (egg) development in the ovary and following ovulation, the production of estrogen. Negative feedback from the estrogen blocks further secretion of FSH. Guys, don't think you needn't worry about FSH: It's present in you, too, encouraging production and maturation of sperm. (For a review of the male and female reproductive systems, flip to Chapters 16 and 17.)

>> **Luteinizing hormone (LH):** Stimulates the production of sex hormones in both males and females and triggers ovulation (release of the egg).

>> **Prolactin (PRL):** Promotes milk production in mammary glands; its role in males is still poorly understood.

>> **Thyrotropic hormone, or thyroid-stimulating hormone (TSH):** Controls the development and release of thyroid gland hormones *thyroxine* (T4) and *triiodothyronine* (T3). The hypothalamus regulates TSH secretion by secreting thyrotropin-releasing hormone (TRH).

>> **Adrenocorticotropic hormone (ACTH), or corticotropin:** Regulates the development, maintenance, and secretions of the cortex of the adrenal gland. Its release is triggered by the hypothalamus in response to stress.

>> **Growth hormone (GH):** Stimulates body weight growth and regulates skeletal growth. This is the only hormone secreted by the anterior lobe that has a general effect on nearly every cell in the body. Its release is triggered by growth hormone releasing hormone (GHRH) from the hypothalamus.

The *posterior lobe*, or *neurohypophysis*, of the pituitary gland stores and releases neurohormone secretions produced by the hypothalamus. This lobe is connected to the hypothalamus by the *hypophyseal tract*, axons with cell bodies lying in the hypothalamus. Whereas the adenohypophysis is made up of epithelial cells, the neurohypophysis is largely composed of nerve fibers and glial cells called *pituicytes*.

Among the hormones produced by the hypothalamus but released from the posterior lobe of the pituitary gland are the following:

>> **Oxytocin (OT):** Stimulates contraction of the uterus during childbirth and the release of breast milk in nursing women. Experts believe that intimate social activities such as hugging, touching, and sexual pleasures release oxytocin in both males and females leading to its role in bonding and its being dubbed "the love hormone."

>> **Vasopressin, or antidiuretic hormone (ADH):** Constricts smooth muscle tissue in the blood vessels, elevating blood pressure and increasing the amount of water reabsorbed by the kidneys, which reduces the production of urine. The hypothalamus has specialized neurons called *osmoreceptors* that monitor the amount of solute in the blood.

See how much of this information you're absorbing:

EXAMPLE

Q. The hormone that stimulates ovulation is the

   a. follicle-stimulating hormone (FSH).

   b. antidiuretic hormone (ADH).

   c. oxytocin (OT).

   d. thyroid-stimulating hormone (TSH).

   e. luteinizing hormone (LH).

A. The correct answer is luteinizing hormone (LH). Don't be fooled into thinking it's FSH; that hormone does its job earlier, when it encourages a follicle to mature.

 8–12 Mark the statement with a T if it's true or an F if it's false:

  8 _____ The pituitary gland consists of two parts: an endocrine gland and modified nerve tissue.

  9 _____ The pituitary gland is found in the sella turcica of the temporal bone.

  10 _____ The adenohypophysis is called the master gland because of its influence on all the body's tissues.

  11 _____ ADH causes constriction of smooth muscle tissue in the blood vessels, which elevates the blood pressure.

  12 _____ The neurohypophysis stores and releases secretions produced by the hypothalamus.

 13 Why is the hypothalamus considered to be just as important as the pituitary gland?

   a. It contracts and relaxes to regulate the so-called master gland.

   b. It brings in substances from the thyroid.

   c. It tells the pituitary gland what to release and when.

   d. It controls how large the other glands can grow.

   e. It influences the other glands more directly than the pituitary does.

---

14 Which of the following is not a pituitary hormone?

   a. Prolactin (PRL)

   b. Follicle-stimulating hormone (FSH)

   c. Growth hormone (GH)

   d. Progesterone (P4)

   e. Luteinizing hormone (LH)

The Supporting Cast of Glandular Characters

While the pituitary orchestrates the show at center stage, the endocrine system enjoys the support of a number of other important glands. Lying in various locations throughout the body, the glands in the following sections secrete check-and-balance hormones that keep the body in tune.

Topping off the kidneys: The adrenal glands

Also called *suprarenals*, the adrenal glands lie atop each kidney like little hats. The central area of each is called the *adrenal medulla*, and the outer layers are called the *adrenal cortex*. Each glandular area secretes different hormones. The cells of the cortex produce more than 30 steroids, including the hormones *aldosterone, cortisol,* and some sex hormones. The medullary cells secrete *epinephrine* (you may know it as *adrenaline*) and *norepinephrine* (also known as *noradrenaline*).

The cortex

Made up of closely packed epithelial cells, the adrenal cortex is loaded with blood vessels. Layers form an outer, middle, and inner zone of the cortex. Each zone is composed of a different cellular arrangement and secretes different steroid hormones.

The following are among the hormones produced by the cortex:

>> *Aldosterone* regulates electrolytes (sodium and potassium mineral salts) retained in the body. It promotes the conservation of water and reduces urine output in the kidneys.

>> *Cortisol* is released in response to stress helping to trigger the "fight-or-flight" response. It acts as an anti-inflammatory agent and can increase blood pressure. It also acts as an antagonist to insulin, increasing blood sugar.

>> *Androgens* and *estrogens* are cortical sex hormones. Androgens generally convey antifemi-nine effects, thus accelerating maleness, although in women adrenal androgens maintain the sexual drive. Too much androgen in females can cause *virilism* (male secondary sexual characteristics). Estrogens have the opposite effect, accelerating femaleness. Too much estrogen in a male produces feminine characteristics.

The medulla

The adrenal medulla is made of irregularly shaped *chromaffin cells* arranged in groups around blood vessels. The sympathetic division of the autonomic nervous system controls these cells as they secrete adrenaline and noradrenaline. The adrenal medulla produces approximately 80 percent adrenaline and 20 percent noradrenaline. Both hormones have similar molecular structure and physiological functions:

>> **Adrenaline** accelerates the heartbeat, stimulates respiration, slows digestion, increases muscle efficiency, and helps muscles resist fatigue.

>> **Noradrenaline** does similar things but also raises blood pressure by stimulating constric-tion of muscular arteries.

The terms "adrenaline" and "noradrenaline" are interchangeable with the terms "epineph-rine" and "norepinephrine." You're likely to encounter both in textbooks and exams.

Thriving with the thyroid

The largest of the endocrine glands, the thyroid is shaped like a large butterfly with two lobes connected by a fleshy *isthmus* positioned in the front of the neck, just below the larynx and on either side of the trachea (you can see the thyroid in Figure 10-1). In response to TSH from the anterior pituitary, the thyroid releases its two primary hormones: *thyroxine* and *triiodothyronine*. These regulate the body's metabolic rate by acting on nearly every cell in the body.

The thyroid produces both T3 and T4, but only T3 is the active form of the hormone. Because they are lipid-soluble, they enter the cell where T4 can be converted to T3 by enzymes. So why does the thyroid even both making T4? It has a longer half-life, meaning it will circulate in the bloodstream for longer than T3.

Parafollicular cells (also called *C cells*) in the thyroid secrete another hormone: *calcitonin*. This hormone promotes bone growth by stimulating osteoblasts (bone-building cells) to take up calcium from the bloodstream and deposit it into the matrix. High blood calcium levels stimu-late the secretion of more calcitonin.

Pairing up with the parathyroid

The parathyroid consists of four pea-sized glands that are attached to the posterior side of the thyroid gland. These secrete *parathyroid hormone* (PTH), which does the opposite of calci-tonin. When blood calcium levels dip, the parathyroid secretes PTH, which increases calcium absorption from the intestine, and triggers osteoclasts to release calcium stored in the bone (by breaking down the matrix). The homeostasis of blood calcium is critical to the conduction of nerve impulses, muscle contraction, and blood clotting.

Pinging the pineal gland

The pineal gland, also called the *epiphysis cerebri*, is a small, oval gland thought to play a role in regulating the body's biological clock. It lies between the cerebral hemispheres and is attached to the thalamus near the roof of the third ventricle.

REMEMBER

Because it both secretes a hormone and receives visual nerve stimuli, the pineal gland is considered part of both the nervous system and the endocrine system. Its hormone *melatonin* is believed to play a role in circadian rhythms, the pattern of repeated behavior associated with the cycles of night and day. The pineal gland is affected by changes in light; when its photo-receptors begin being stimulated less and less, it secretes melatonin, making you feel sleepy.

Thumping the thymus

The thymus secretes a peptide hormone called *thymosin,* which promotes the production and maturation of T lymphocyte cells. T cells are a type of white blood cell that play a key role in immunity — mounting a targeted attack against the invading germ. This is discussed further in Chapter 12. The gland is large in children and atrophies with age because the "training" of T cells is mostly complete by the onset of puberty.

Pressing the pancreas

The pancreas is both an exocrine and an endocrine gland, which means that it secretes some substances through ducts while others go directly into the bloodstream. (We cover its exocrine functions in Chapter 14.) The pancreatic endocrine glands are clusters of cells called the *islets of Langerhans.* Within the islets are a variety of cells, including

>> **A cells (alpha cells)** that secrete the hormone *glucagon* to increase blood sugar by triggering *glycogen* breakdown in the liver (glycogen is the storage form of glucose).

>> **B cells (beta cells)** that secrete *insulin* to decrease blood sugar levels by triggering the liver to store glucose, increases lipid synthesis, and stimulates protein synthesis.

>> **D cells (delta cells)** that secrete *somatostatin,* to inhibit the secretion of insulin and glucagon.

>> **F cells (PP cells)** that secrete a pancreatic polypeptide that regulates the release of pancreatic digestive enzymes.

WARNING

We tend to think of insulin as a necessity to keep blood sugar level down in the normal range, which it does. But its mechanism in doing so is more important. When insulin binds to receptors on a cell's surface, it triggers the opening of glucose channels, allowing glucose to diffuse into the cell. Without insulin, cells cannot obtain enough glucose for cellular respiration and thus cannot produce enough ATP to carry out their functions.

See whether all this information has your hormones raging:

15-19 Mark the statement with a T if it's true or an F if it's false:

15 _____ The adrenal glands are located in the cortex of the kidneys.

16 _____ Adrenaline is functional in the absorption of stored carbohydrates and fat.

17 _____ Aldosterone is functional in regulating the amount of insulin in the body.

18 _____ The sympathetic division of the autonomic nervous system controls the cells of the adrenal medulla.

19 _____ The adrenal cortex produces sex hormones.

20 The thymus produces thymosin, which

a. stimulates the pineal gland.

b. inhibits the pituitary gland.

c. initiates lymphocyte development.

d. stimulates the thyroid gland.

e. controls the hypothalamus.

21-24 Mark the statement with a T if it's true or an F if it's false.

21 _____ Parafollicular cells of the thyroid produce hormones that affect the metabolic rate of the body.

22 _____ Thyroxine (T4) is normally secreted in lower quantity than triiodothyronine (T3).

23 _____ The hormone calcitonin helps regulate the concentration of sodium and potassium.

24 _____ The parathyroid gland contains cells that secrete parathyroid hormone (PTH), which works to increase blood calcium levels.

25 Which statement is NOT true of the pineal gland?

a. It secretes melatonin.

b. Nerve fibers stimulate the pineal cells.

c. As light decreases, secretion increases.

d. It promotes immunity.

e. It's considered part of both the endocrine and nervous systems.

26 The adrenal medulla serves one primary function. What is it?

   a. To produce erythrocyte cells

   b. To absorb excess iodine from the bloodstream

   c. To produce epinephrine

   d. To secrete polypeptides

   e. To regulate water retention in the kidneys

27 Clusters of cells can be found within the islets of Langerhans in the pancreas. Which variety of cell secretes glucagon to increase blood sugar?

   a. Beta cells, also known as B cells

   b. Alpha cells, also known as A cells

   c. PP cells, also known as F cells

   d. Delta cells, also known as D cells

28-37 Use the terms that follow to identify the structures of the endocrine system shown in Figure 10-2.

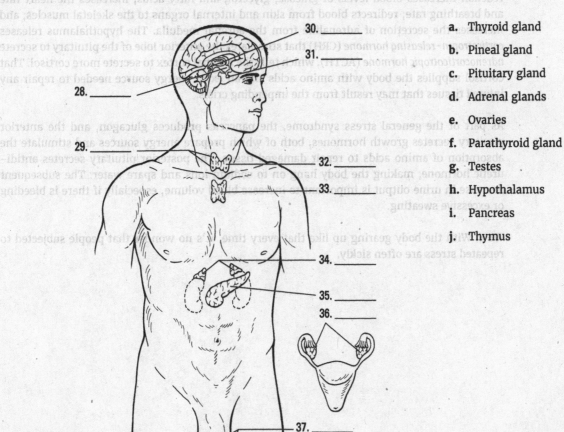

28. _____
29. _____
30. _____
31. _____
32. _____
33. _____
34. _____
35. _____
36. _____
37. _____

   a. Thyroid gland

   b. Pineal gland

   c. Pituitary gland

   d. Adrenal glands

   e. Ovaries

   f. Parathyroid gland

   g. Testes

   h. Hypothalamus

   i. Pancreas

   j. Thymus

FIGURE 10-2:
The endocrine system.

Illustration by Kathryn Born, MA

Dealing with Stress: Homeostasis

Nothing upsets your delicate cells more than a change in their internal environment. A stimulus such as fear or pain provokes a response that upsets your body's carefully maintained equilibrium. Such a change initiates a nerve impulse to the hypothalamus that activates the sympathetic division of the autonomic nervous system and increases secretions from the adrenal glands. This change — called a *stressor* — produces a condition many know all too well: *stress*. Stress is anything that disrupts the normal function of the body. The body's immediate response is to push for *homeostasis* — keeping everything the same inside. Inability to maintain homeostasis — a condition known as *homeostatic imbalance* — may lead to disease or even death.

The body's effort to maintain homeostasis invokes a series of reactions called the *general adaptation syndrome*, the predictable phases of response to stress that's initiated by the hypothalamus. When the hypothalamus receives stress information, it responds by preparing the body for fight or flight — in other words, some kind of decisive, immediate, physical action. This reaction increases blood levels of glucose, glycerol, and fatty acids; increases the heart rate and breathing rate; redirects blood from skin and internal organs to the skeletal muscles; and increases the secretion of adrenaline from the adrenal medulla. The hypothalamus releases *corticotropin-releasing hormone* (CRH) that stimulates the anterior lobe of the pituitary to secrete *adrenocorticotropic hormone* (ACTH), which tells the adrenal cortex to secrete more cortisol. That cortisol supplies the body with amino acids and an extra energy source needed to repair any injured tissues that may result from the impending crisis.

As part of the general stress syndrome, the pancreas produces glucagon, and the anterior pituitary secretes growth hormones, both of which prepare energy sources and stimulate the absorption of amino acids to repair damaged tissue. The posterior pituitary secretes antidiuretic hormone, making the body hang on to sodium ions and spare water. The subsequent decrease in urine output is important to increase blood volume, especially if there is bleeding or excessive sweating.

Wow. With the body gearing up like that every time, it's no wonder that people subjected to repeated stress are often sickly.

We'll try not to stress you out with these practice questions:

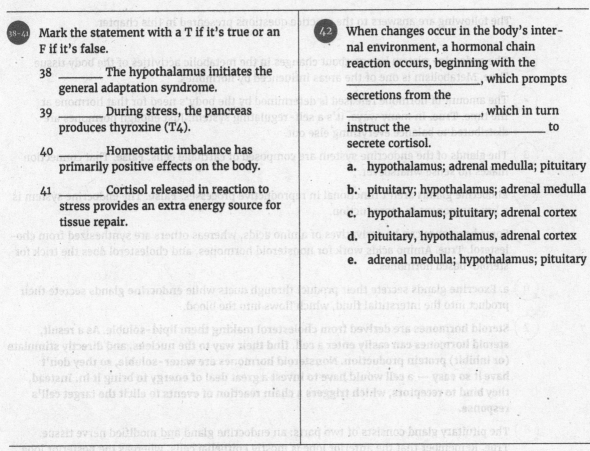

38-41 Mark the statement with a T if it's true or an F if it's false.

38 _____ The hypothalamus initiates the general adaptation syndrome.

39 _____ During stress, the pancreas produces thyroxine (T4).

40 _____ Homeostatic imbalance has primarily positive effects on the body.

41 _____ Cortisol released in reaction to stress provides an extra energy source for tissue repair.

42 When changes occur in the body's internal environment, a hormonal chain reaction occurs, beginning with the _____, which prompts secretions from the _____, which in turn instruct the _____ to secrete cortisol.

a. hypothalamus; adrenal medulla; pituitary

b. pituitary; hypothalamus; adrenal medulla

c. hypothalamus; pituitary; adrenal cortex

d. pituitary, hypothalamus, adrenal cortex

e. adrenal medulla; hypothalamus; pituitary

43 The body's initial reaction to a stressor is

a. the fight-or-flight response.

b. the repair response.

c. to promote rapid wound healing.

d. the stress reflex.

e. to promote normal metabolism.

Answers to Questions on the Endocrine System

The following are answers to the practice questions presented in this chapter.

1. The endocrine system brings about changes in the metabolic activities of the body tissue. **True.** Metabolism is one of the areas influenced by hormones.

2. The amount of hormone released is determined by the body's need for that hormone at the time. **True.** In many ways, it's a self-regulating system: Just enough hormones are distributed to balance everything else out.

3. The glands of the endocrine system are composed of cartilage cells. **False.** That connection makes no sense whatsoever.

4. Endocrine glands aren't functional in reproductive processes. **False.** The endocrine system is a key component in reproduction.

5. Some hormones can be derivatives of amino acids, whereas others are synthesized from cholesterol. **True.** Amino acids work for nonsteroid hormones, and cholesterol does the trick for steroid-based hormones.

6. **a. Exocrine** glands secrete their product through ducts while **endocrine** glands secrete their product into the interstitial fluid, which flows into the blood.

7. **Steroid hormones are derived from cholesterol making them lipid-soluble. As a result, steroid hormones can easily enter a cell, find their way to the nucleus, and directly stimulate (or inhibit) protein production. Nonsteroid hormones are water-soluble, so they don't have it so easy — a cell would have to invest a great deal of energy to bring it in. Instead, they bind to receptors, which triggers a chain reaction of events to elicit the target cell's response.**

8. The pituitary gland consists of two parts: an endocrine gland and modified nerve tissue. **True.** Remember that the anterior lobe is mostly epithelial cells, whereas the posterior lobe contains primarily nerves (bundled axons).

9. The pituitary gland is found in the sella turcica of the temporal bone. **False.** The pituitary gland is found in the sella turcica but that's part of the sphenoid bone, not the temporal bone.

10. The adenohypophysis is called the master gland because of its influence on all the body's tissues. **False.** It earned the title "master gland" because of its influence over the other endocrine glands.

11. ADH causes constriction of smooth muscle tissue in the blood vessels, which elevates the blood pressure. **True.** It's understandable that constriction increases pressure. This, however, is a secondary function to its influence on the kidneys to retain water.

12. The neurohypophysis stores and releases secretions produced by the hypothalamus. **True.** Seems a strange thing for a structure made of nerves to do, but it does its job well.

13. Why is the hypothalamus considered to be just as important as the pituitary gland? **c. It tells the pituitary gland what to release and when.** If you remember that the pituitary gland is attached to the bottom of the hypothalamus, you can see the importance of the relationship between the two.

(14) Which of the following is not a pituitary hormone? **d. Progesterone.** That's made by the corpus luteum in the ovary following ovulation.

(15) The adrenal glands are located in the cortex of the kidneys. **False.** They're atop the kidneys.

(16) Adrenaline is functional in the absorption of stored carbohydrates and fat. **False.** Adrenaline does lots of things, but not that.

(17) Aldosterone is functional in regulating the amount of insulin in the body. **False.** This hormone regulates the uptake of mineral salts and thus the conservation of water by the kidneys.

(18) The sympathetic division of the autonomic nervous system controls the cells of the adrenal medulla. **True.** Those cells are busy secreting both adrenaline and noradrenaline.

(19) The adrenal cortex produces sex hormones. **True.** Production of these hormones is not limited to the gonads.

(20) The thymus produces thymosin, which **c. initiates lymphocyte development.** Specifically, the production and training of T cells.

(21) Parafollicular cells of the thyroid produce hormones that affect the metabolic rate of the body. **False.** These are the cells that produce calcitonin.

(22) Thyroxine (T4) is normally secreted in a lower quantity than triiodothyronine (T3). **False.** In fact, it's just the opposite — more T4 is secreted than T3 because it lasts longer.

(23) The hormone calcitonin helps regulate the concentration of sodium and potassium. **False.** Actually, calcitonin lowers plasma calcium and phosphate levels.

(24) The parathyroid gland contains cells that secrete parathyroid hormone (PTH) which works to increase blood calcium levels. **True.** As far as we know, PTH is the only hormone that can trigger the breakdown of bone matrix to release the calcium stored there.

(25) Which statement is NOT true of the pineal gland? **d. It promotes immunity.** It's more of a biological-clock kind of gland.

(26) The adrenal medulla serves one primary function. What is it? **c. To produce epinephrine.** That's about all the adrenal medulla does. (Actually, it produces 80 percent epinephrine and 20 percent norepinephrine.)

(27) Clusters of cells can be found within the islets of Langerhans in the pancreas. Which variety of cell secretes glucagon to increase blood sugar? **b. Alpha cells, also known as A cells.** The secretion from A cells increases blood sugar, insulin from B cells decreases blood sugar, somatostatin from D cells inhibits both insulin and glucagon, and the polypeptide from F cells regulates digestive enzymes.

(28-37) Following is how Figure 10-2, the endocrine system, should be labeled.

28. h. Hypothalamus; 29. f. Parathyroid gland; 30. b. Pineal gland; 31. c. Pituitary gland; 32. a. Thyroid gland; 33. j. Thymus; 34. d. Adrenal glands; 35. i. Pancreas; 36. e. Ovaries; 37. g. Testes.

(38) The hypothalamus initiates the general adaptation syndrome. **True.** It does so by initiating the body's fight or flight response.

39 During stress, the pancreas produces thyroxine (T4). **False.** That's the thyroid's job, and T4 has nothing to do with stress, anyway.

40 Homeostatic imbalance has primarily positive effects on the body. **False.** Imbalance is a bad thing — so bad that it's capable of killing.

41 Cortisol released in reaction to stress provides an extra energy source for tissue repair. **True.** It's part of how the body gears up to manage an impending crisis.

42 When changes occur in the body's internal environment, a hormonal chain reaction occurs, beginning with the **c. hypothalamus,** which prompts secretions from the **pituitary,** which instruct the **adrenal cortex** to secrete cortisol.

43 The body's initial reaction to a stressor is **a. the fight-or-flight response.** You're either going to put up your dukes or run like the wind — or freeze, as some psychologists argue that it should be renamed the fight-flight-or-freeze response.

4

Feed and Fuel: Supply and Transport

Take the pulse of the cardiovascular system and discover how its internal transit routes carry nutrients and oxygen to every nook and cranny of the body.

Delve into the lymphatic ducts and find out about the body's internal defense forces.

Breathe in the details of the respiratory system, including the parts of the respiratory tract (the nose, the sinuses, the throat, and the lungs) and how oxygen finds its way into the bloodstream.

Fuel up with food and then study what happens to it in the digestive system after every possible nutrient has been wrung from it.

Bat cleanup with the kidneys and track how they remove waste from the body as part of the urinary system.

Chapter **11**

Spreading the Love: The Cardiovascular System

This chapter gets to the heart of the well-oiled human machine to see how its central pump is the hardest-working muscle in the entire body. From a month after you're conceived to the moment of your death, this phenomenal powerhouse pushes a liquid connective tissue — blood — and its precious cargo of oxygen and nutrients to every nook and cranny of the body, and then it keeps things moving to bring carbon dioxide and waste products back out again. In the first seven decades of human life, the heart beats roughly 2.5 billion times. Do the math: How many pulses has your ticker clocked if the average heart keeps up a pace of 72 beats per minute, 100,000 per day, or roughly 36 million per year?

Moving to the Beat of a Pump

The cardiovascular system includes the heart, all the blood vessels, and the blood that flows through them. While it is still sometimes referred to as the *circulatory system*, that terminology is falling out of favor as our body circulates another fluid, called lymph, through a different system. (Chapter 12 discusses the lymphatic system.) The cardiovascular system functions as a closed double loop. Closed because the blood never leaves the vessels; double because there are two distinct circuits (see Figure 11-1).

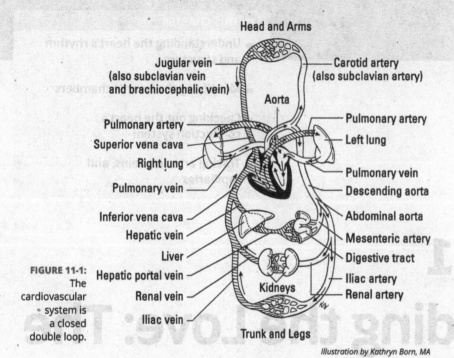

Head and Arms

Jugular vein
(also subclavian vein
and brachiocephalic vein)

Carotid artery
(also subclavian artery)

Aorta

Pulmonary artery
Superior vena cava
Right lung
Pulmonary vein

Pulmonary artery
Left lung
Pulmonary vein
Descending aorta

Inferior vena cava
Hepatic vein
Liver
Hepatic portal vein
Renal vein
Iliac vein

Abdominal aorta
Mesenteric artery
Digestive tract
Iliac artery
Renal artery

Kidneys

Trunk and Legs

FIGURE 11-1:
The
cardiovascular
system is
a closed
double loop.

Illustration by Kathryn Born, MA

The heart is separated into right and left halves by the septum, allowing blood to flow through each half without mixing. With each *cardiac cycle* (a single set of muscle contractions), blood moves through the right and left sides simultaneously, thus creating the two circuits — the right side destined for the lungs and the left to, well, everywhere else.

These two circuits have distinct functions:

WARNING

>> **Pulmonary:** The *pulmonary circuit* carries blood to and from the lungs for gas exchange. Deoxygenated blood saturated with carbon dioxide enters the right side of the heart from the *inferior* and *superior vena cavae*. From there, it is pumped out through the *pulmonary trunk* to *pulmonary arteries* to capillaries in the lungs, where the carbon dioxide departs the system. That same blood, freshly loaded with oxygen *(oxygenated)*, then returns to the left side of the heart through the *pulmonary veins*, where it enters the second circuit.

Though we tend to associate arteries with oxygenated blood and veins with deoxygenated blood, the terms refer to their structure. Arteries are adapted to carry blood away from the heart under high pressure, while veins are the opposite. As such, the pulmonary arteries carry deoxygenated blood and the pulmonary veins carry oxygenated blood.

>> **Systemic:** The *systemic circuit* uses the oxygenated blood to maintain a constant internal pressure and flow rate around the body's tissues, allowing them access to the oxygen and other vital nutrients in the blood. The oxygenated blood enters the left side of the heart through the *pulmonary veins* and is pumped out through the *aorta* to be distributed throughout the body. Branching off the aorta is the *coronary artery*, creating a side circuit to provide

resources to the *myocardium* (the muscular wall of the heart). From here, the *cardiac veins* carry the deoxygenated blood to the right side. Blood from the rest of the body is brought back to the heart in the vena cavae.

Although cutely depicted in popular culture as uniformly curvaceous, the heart actually looks more like a blunt, muscular cone resting on the diaphragm. The sternum (breastbone) and third to sixth costal cartilages of the ribs provide protection in front of (ventrally to) the heart. Behind it lie the fifth to eighth thoracic vertebrae. (Check out Chapter 7 to review the skeleton.) Two-thirds of the heart lies to the left of the body's center nestled into the left lung.

While in the body, the heart is not positioned as you see in diagrams. It is twisted slightly so the right side is anterior (more forward) to the left side. The *apex* is the point of the heart and the *base* is the opposite end where all the blood vessels attach.

So, the base is the top and the apex is the bottom? Isn't that backwards? Well, yes, sort of. But as you'll see by the time you finish this chapter, contraction starts at the apex to push the blood toward the base. If you think of it that way, you won't get the two confused.

Three layers make up the wall of the heart:

>> **Epicardium:** On the outside lies the *epicardium,* which is composed of connective tissue containing adipose tissue (fat) to protect the underlying muscle. It forms a framework for nerves as well as the blood vessels that supply nutrients to the heart's tissues.

>> **Myocardium:** Beneath the epicardium lies the *myocardium,* which is composed of organized layers and bundles of cardiac muscle tissue.

>> **Endocardium:** The *endocardium,* the heart's interior lining, is composed of simple squamous endothelial cells. This layer is continuous with all of the systems' blood vessels.

Too much to remember? To keep the layers straight, turn to the Greek roots. *Epi–* is the Greek term for "upon" or "on" whereas *endo–* comes from the Greek *endon* meaning "within." The medical definition of *myo–* is "muscle." And *peri–* comes from the Greek term for "around" or "surround." Hence the *epi*cardium is *on* the heart, the *endo*cardium is *inside* the heart, the *myo*cardium is the *muscle* between the two, and the *peri*cardium surrounds the whole package. By the way, the root *cardi–* comes from the Greek word for heart, *kardia.*

Since the heart beats with such gusto it's isolated from all of the other structures in the mediastinum in a protective sac called the *pericardium*. The pericardium is composed of multiple layers of connective tissue. The outermost layers comprise the *parietal pericardium,* and the innermost is the *visceral pericardium* (which is also known as the epicardium). Space is left between these layers creating the *pericardial cavity*. This space, albeit tiny, is filled with a serous fluid to reduce friction, preventing damage to the heart's walls as they are moved with each set of contractions.

Have you got the beat? Let's try a few practice questions:

 1-5 Match the description to its anatomical term.

   a. Pericardium

   b. Pulmonary circuit

   c. Systemic circuit

   d. Epicardium

   e. Septum

   1 _____ The system for gaseous exchange in the lungs

   2 _____ The system for maintaining a constant internal environment in other tissues

   3 _____ The membranous sac that surrounds the heart

   4 _____ The wall that divides the heart into two cavities

   5 _____ The layer providing the heart's wall with nutrients

 6 What's the difference between the pulmonary and systemic circuits?

   a. The pulmonary circuit supplies blood to the brain, and the systemic circuit supplies it everywhere else.

   b. The pulmonary circuit supplies blood to the extremities, and the systemic circuit supplies blood to the internal organs.

   c. The pulmonary circuit supplies blood to the skin, and the systemic circuit supplies blood to the underlying muscles.

   d. The pulmonary circuit supplies blood to the left side of the body, and the systemic circuit supplies blood to the right side.

   e. The pulmonary circuit travels to and from the lungs, and the systemic circuit keeps pressure and flow up for the body's tissues.

 7 True or False: The heart is centrally located in the chest.

 8 Which blood vessel contains oxygenated blood?

   a. pulmonary artery

   b. pulmonary vein

   c. pulmonary trunk

   d. inferior vena cava

   e. superior vena cava

9 Only one of the heart wall's three layers contains muscle tissue. Which one?

    a. Pericardium

    b. Epicardium

    c. Myocardium

    d. Endocardium

    e. Endothelium

Finding the Key to the Heart's Chambers

Each cavity of the heart is divided into two chambers: the right and left atria on top and the right and left ventricles on bottom. The ventricles are quite a bit larger than the two atria up top. Yet the proper anatomical terms for their positions refer to the atria as being "superior" (above) and the ventricles "inferior" (below). In this case, size isn't the issue at all. Between each atrium and ventricle is an *atrioventricular valve (AV valve)*, which have leaflets, or *cusps*, that open and close to control entry into the ventricles. The exits of the ventricles (into the arteries) are guarded by *semilunar valves*. The three cups that form these valves are half-moon shaped, hence the name.

The atria

The atria are sometimes referred to as "receiving chambers" because they receive blood returning to the heart through the veins. Each atrium has two parts: a *principal cavity* with a smooth interior surface and an *auricle* (atrial appendage), a smaller, dog-ear-shaped pouch with muscular ridges inside called *pectinate muscles*. The atria have thin walls as they only need to generate enough force to pump blood into the ventricles. The right atrium is slightly larger than the left and the *interatrial septum* lies between them.

Deoxygenated blood passively (pressure is lower in the atrium than in the blood vessels) fills the right atrium through three openings:

>> The *superior vena cava,* which has no valve and returns blood from the head, thorax, and upper extremities.

>> The *inferior vena cava,* which returns blood from the trunk and lower extremities.

>> The *coronary sinus,* returns blood from the heart, and is covered by the ineffective *Thebesian valve.*

At the bottom of the right atrium is the tricuspid valve. Its three flaps open when the pressure in the right ventricle drops below that of the right atrium. This allows the blood to flow into the right ventricle. Attached to the tricuspid valve on the ventricular side (underneath) are strong strings called *chordae tendineae*. These fuse with *papillary muscles* in the walls of the ventricle. The chordae tendineae are pulled tight when the ventricle contracts to hold the tricuspid valve closed, preventing the backflow of blood.

The left atrium's principal cavity contains openings for the four *pulmonary veins*, two from each lung, which have no valves. Frequently, the two left veins share a common opening. The left auricle, a dog-ear-shaped blind pouch, is longer, narrower, and more curved than the right. The left atrium's AV opening is smaller than on the right and is guarded by the *mitral*, or *bicuspid, valve*. Though it has only two cusps, it contains chordae tendineae and functions the same as the tricuspid valve.

The ventricles

The heart's ventricles are sometimes called the pumping chambers because it's their job to receive blood from the atria and pump it out through the arteries. More force is needed to move the blood great distances, so the myocardium of the ventricles is thicker than that of either atrium, and the myocardium of the left ventricle is thicker than that of the right. (It takes greater pressure to maintain the systemic circuit.) The two ventricles are separated by a thick wall of muscle called the *interventricular septum*.

The right ventricle only has to move blood to the lungs, so its myocardium is only one-third as thick as that of its neighbor to the left. Roughly triangular in shape, the right ventricle occupies much of the *sternocostal* (front) surface of the heart and joins the *pulmonary trunk* just left of the AV opening. The right ventricle extends downward toward where the heart rests against the diaphragm. Cardiac muscle in the ventricle's wall is in an irregular pattern of bundles and bands called the *trabeculae carneae*. The circular opening into the pulmonary trunk is covered by the *pulmonary valve* that opens when the ventricle contracts (though not immediately). When the ventricle relaxes, the blood from the pulmonary artery tends to flow back toward the ventricle, filling the pockets of the cups and causing the valve to close. Openings into the ventricles are surrounded by strong, fibrous rings sometimes referred to as the *cardiac skeleton* (though there are definitely no bones here).

Longer and more conical in shape, the left ventricle's tip forms the apex of the heart. More ridges are packed more densely in the muscular trabeculae carneae. Its opening to the aorta is protected by the *aortic valve* that is larger, thicker, and stronger than the pulmonary valve's cups. The blood enters the ascending aorta into the aortic arch where three vessels branch off:

>> The brachiocephalic trunk that will divide into the right common carotid artery and the right subclavian artery

>> The left common carotid artery

>> The left subclavian artery

Pump up your practice time with these questions related to the chambers of the heart:

10-25 Use the terms that follow to identify the heart's major vessels shown in Figure 11-2.

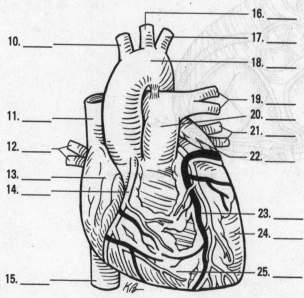

FIGURE 11-2: The heart and major vessels.

Illustration by Kathryn Born, MA

a. Left pulmonary veins

b. Left ventricle

c. Brachiocephalic trunk

d. Right pulmonary veins

e. Right ventricle

f. Left subclavian artery

g. Right coronary artery

h. Aortic arch

i. Great (left) cardiac vein

j. Superior vena cava

k. Left common carotid artery

l. Left pulmonary arteries

m. Left atrium

n. Inferior vena cava

o. Pulmonary trunk

p. Right atrium

26-29 Use the terms that follow to identify the heart valves shown in Figure 11-3.

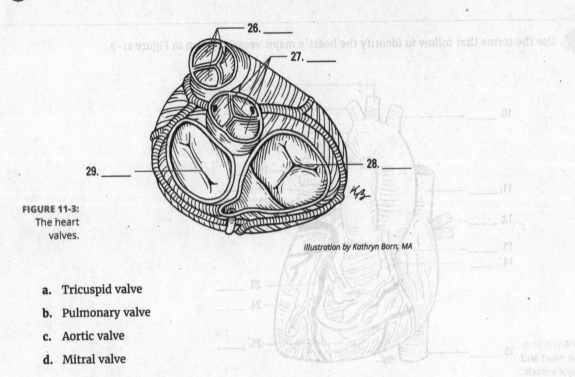

26. _____

27. _____

29. _____

28. _____

FIGURE 11-3:
The heart
valves.

Illustration by Kathryn Born, MA

a. Tricuspid valve

b. Pulmonary valve

c. Aortic valve

d. Mitral valve

---

30 Why is the myocardium of the ventricles thicker than the myocardium of the atria?

a. The myocardium keeps valves in place, and the ventricular valves have to hold back greater blood pressure with each contraction.

b. The myocardium is the inner lining of the heart, and the ventricles must hold greater volumes of blood between contractions.

c. The myocardium is the muscular tissue of the heart, and more force is needed in chambers pumping blood greater distances.

d. The myocardium consists mostly of connective tissue, and the ventricles are more intricate than the atria.

31 The superior vena cava enters the heart by way of the

a. left ventricle.

b. right ventricle.

c. left atrium.

d. right atrium.

32 What holds the cusps of the atrioventricular valves in place?

   a. Supporting ligaments

   b. The chordae tendineae

   c. The trabeculae carneae

   d. Pectinate muscles

   e. The papillary muscles

33 What covers the atrioventricular opening between the right atrium and the right ventricle?

   a. Bicuspid valve

   b. Tricuspid valve

   c. Pulmonary valve

   d. Mitral valve

   e. Aortic valve

34–40 Match the following descriptions with the proper anatomical terms.

   a. Tricuspid valve

   b. Mitral valve

   c. Superior vena cava

   d. Pulmonary valve

   e. Inferior vena cava

   f. Aortic valve

   g. Pulmonary vein

34 ___C___ Returns blood to the heart from the head, thorax, and upper extremities

35 ___A___ Valve located between the right atrium and right ventricle

36 ___D___ Valve located between the right ventricle and pulmonary trunk

37 ___E___ Returns blood to the heart from the trunk and lower extremities

38 ___B___ Valve located between the left atrium and left ventricle

39 ___F___ Semilunar valve located in the left ventricle

40 ___G___ Returns blood from the lungs

Conducting the Heart's Music

The mighty, nonstop heart keeps up its rhythm because of a carefully choreographed dance of electrical impulses called the *cardiac conduction system* that has the power to produce a spontaneous rhythm and conduct an electrical impulse. The feature unique to cardiac muscle, the presence of *intercalated discs*, plays an essential role here. These undulating double membranes separate adjacent cardiac muscle fibers with *gap junctions* permitting ions to pass between the cells. This allows the spread of the *action potential* of the electrical impulse throughout the myocardium, synchronizing cardiac muscle contractions.

The orchestra

Five structures play key roles in this dance — the *sinoatrial node*, *atrioventricular node*, *atrioventricular bundle (bundle of His)*, *bundle branches*, and *Purkinje fibers*. (You can see them in Figure 11-4.) Each is formed of highly tuned, modified cardiac muscle. Rather than both contracting and conducting impulses as other cardiac muscle does, these structures specialize in conduction alone, setting the pace for the rest of the heart. Following is a bit more information about each one:

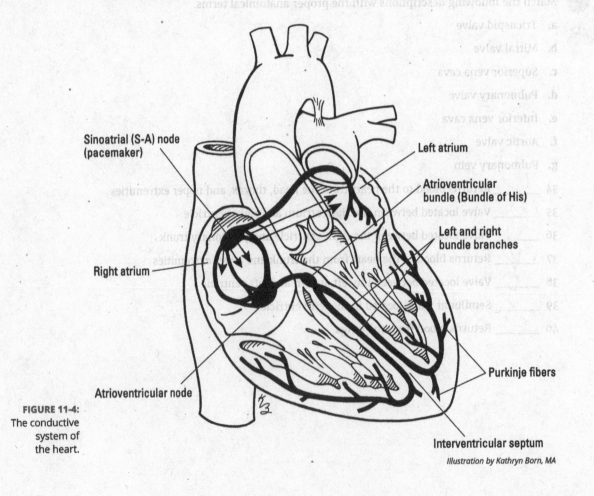

FIGURE 11-4: The conductive system of the heart.

Illustration by Kathryn Born, MA

REMEMBER

>> **Sinoatrial node (SA node):** This node really is the pacemaker of the heart. Located at the junction of the superior vena cava and the right atrium, this small knot, or mass, of specialized cardiac muscle cells initiates an electrical impulse that moves over the musculature of both atria. This network of conducting fibers forms the *atrial syncytium,* causing atrial walls to contract simultaneously and emptying blood into both ventricles.

>> **Atrioventricular node (AV node):** The impulse that starts in the SA node moves to this mass of conducting fibers that's located at the junction of all four chambers of the heart. Fibers carrying impulse from the SA node to the AV node are smaller, slowing it down to ensure that the ventricles contract after the atria.

>> **Atrioventricular bundle:** From the AV node, the impulse moves into the atrioventricular bundle, also known as the *AV bundle* or *bundle of His* (pronounced "hiss").

>> **Bundle branches:** The atrioventricular bundle breaks into two branches that extend down the sides of the interventricular septum under the endocardium to the heart's apex.

>> **Purkinje fibers:** At the apex, the bundles branch into Purkinje fibers, which spread throughout the myocardium forming the *ventricular syncytium.* Both ventricles then contract together in an upward, twisting motion beginning at the apex and moving toward the base of the heart, forcing blood into the aorta and pulmonary trunk.

The music

Though we can stimulate the SA node through the autonomic nervous system, it is self-stimulating, so we need only do so to alter the normal, or *sinus rhythm.* While we tend to think of contraction as a quick event, it's more like a phase. When the chambers are contracting, they are said to be in *systole* (pronounced *sis-toe-lee*); when relaxed they are in *diastole* (pronounced *di-as-toe-lee*). To begin the cardiac cycle, all chambers are in systole and all valves are closed. It proceeds as follows:

1. Blood trickles into the atria from the veins, increasing the internal pressure.

2. When the pressure in the atria rises above that of the ventricles (they would be relaxing at this time thus dropping the pressure) the AV valves are pushed open.

3. Blood passively fills the ventricles until they are about 70-percent full. At this point, the pressures of the chambers have equalized.

4. Atrial systole begins, pushing the remaining 30 percent of the blood into the ventricles.

5. The atria begin to relax as the ventricles now begin systole. This drops the pressure of the atrium creating a vacuum. The cusps of the AV valves are thus pulled upwards. However, the papillary muscles are contracting, pulling on the chordae tendineae, preventing the cusps from being pulled into the atrium.

6. At this point, all valves are closed, the atria are in diastole, and the ventricles are in systole.

7. The ventricles continue contracting to increase the internal pressure. Once it rises above the pressure inside the arteries, the semilunar valves are pushed open.

8. Blood enters the arteries to be carried elsewhere and the ventricles begin diastole.

9. Again, this creates a vacuum pulling the blood backwards, this time toward the ventricles.

10. Blood catches in the cups of the semilunar valves, sealing them shut (there are no chordae tendineae here). And now we're back to the beginning.

WARNING It's easy to fall into the trap of thinking that contractions open and close the valves. However, this is not so. With the exception of the AV valves closing (which is initiated by a pressure change but maintained by ventricular contraction), all valve movements are triggered by pressure changes.

The performance

The cardiac cycle generates the forces necessary to propel the blood along the two circuits. The spread of impulse is measurable through an electrocardiogram (EKG or ECG) which generates the familiar spikes visible on a heart rate monitor. Keep in mind, though, that these spikes precede the contractions and are not the contractions themselves.

Similarly, when listening to the heartbeat, the sounds you hear are not the contractions. A healthy heart makes a "lub-dub" sound as it beats because a single cardiac cycle contains two separate contractions: The first sound (the "lub") is heard most clearly near the apex of the heart and comes at the beginning of ventricular systole — the sound you hear is the atrioventricular valves snapping shut. It's lower in pitch and longer in duration than the second sound (the "dub"), heard most clearly over the second rib, which results from the semilunar valves closing during ventricular diastole.

Give these practice questions a try to see if you've got the rhythm of all of this:

EXAMPLE

Q. Cardiac tissue is distinctive microscopically because of the presence of

a. hemoglobin.

b. intercalated discs.

c. fibrin.

d. myofibrils

e. ganglia.

A. The correct answer is intercalated discs. These undulating double membranes separate adjacent cardiac muscle fibers, so they're unique to cardiac tissue.

41   Why is the sinoatrial node (SA node) called the heart's "pacemaker"?

    a. It interrupts irregular heart rhythms to set a steadier pace.

    b. It initiates an electrical impulse that causes atrial walls to contract simultaneously.

    c. It speeds up and slows down the heart, depending on the body's activities.

    d. It tells the ventricles to contract simultaneously.

    e. It supplies a boost of energy to jump-start the heart when it slows.

42   True or False: The chordae tendineae open the AV valves.

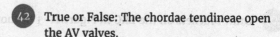

43   Choose the correct conductive pattern.

    a. S–A node → bundle of His → bundle branches → A–V node → Purkinje fibers

    b. A–V node → S–A node → Purkinje fibers → bundle of His → bundle branches

    c. S–A node → A–V node → bundle of His → bundle branches → Purkinje fibers

    d. S–A node → Purkinje fibers → bundle of His → bundle branches → A–V node

44   Which valve action results from the pressure in the ventricle being lower than that of the atria?

    a. semilunar valves open

    b. semilunar valves close

    c. AV valves open

    d. AV valves close

45 Draw brackets on the EKG in Figure 11-5 to show atrial and ventricular systole.

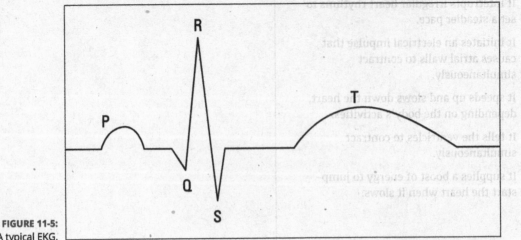

FIGURE 11-5:
A typical EKG.

Illustration by Kathryn Born, MA

46 True or False: The atrial syncytium is comprised of the conducting fibers of the atrium that ensure the atria contract before the ventricles.

47 Starting with a drop of blood returning to the heart from the legs, trace its path until its eventual return to the legs.

_____

_____

_____

_____

_____

_____

_____

_____

Riding the Network of Blood Vessels

REMEMBER

Blood vessels come in three varieties, which you can see illustrated in Figure 11-6:

>> **Arteries:** *Arteries* carry blood away from the heart; all arteries, except the pulmonary arteries, carry oxygenated blood. The largest artery is the aorta. Small ones are called *arterioles*, and these send off branches called *metarterioles* that join with capillaries.

>> **Veins:** *Veins* carry blood toward the heart; all veins except the pulmonary veins contain deoxygenated blood. Small ones are called *venules*, into which the capillaries drain.

>> **Capillaries:** Microscopically small *capillaries* carry blood from arterioles to venules. These thin-walled vessels are the site of *capillary exchange*, where tissues receives resources from the blood.

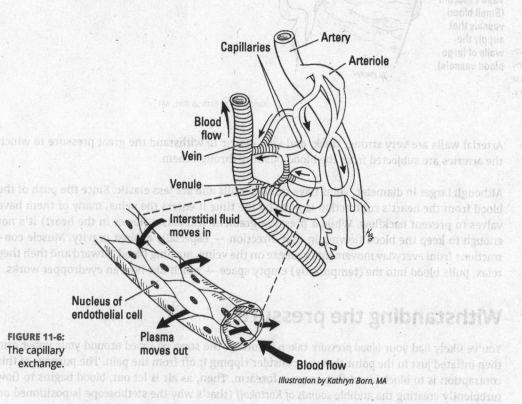

FIGURE 11-6: The capillary exchange.

Illustration by Kathryn Born, MA

The walls of arteries and veins have three layers (see Figure 11-7):

>> The outermost *tunica externa* with numerous collagen fibers to anchor the vessel to surrounding structures composed of connective tissue.

>> A central "active" layer called the *tunica media* composed of smooth muscle fibers and elastic connective tissue.

>> An inner layer called the *tunica intima* made up of simple squamous endothelium is continuous through all blood vessels. Cells of this layer secrete chemicals to prevent blood coagulation. Additionally, they regulate blood flow by influencing level of contraction in the tunica media.

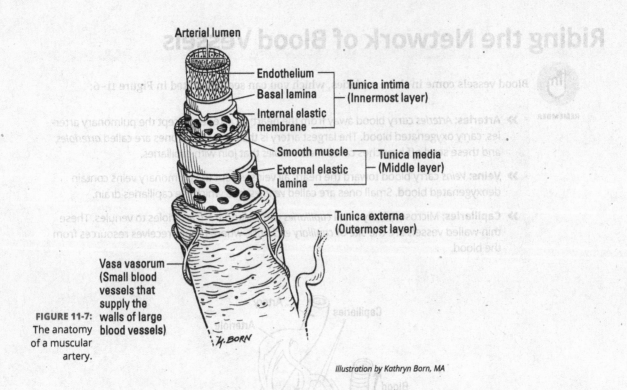

Arterial lumen

Endothelium
Basal lamina
Internal elastic
membrane

Tunica intima
(Innermost layer)

Smooth muscle
External elastic
lamina

Tunica media
(Middle layer)

Tunica externa
(Outermost layer)

Vasa vasorum
(Small blood
vessels that
supply the
walls of large
blood vessels)

FIGURE 11-7:
The anatomy
of a muscular
artery.

Illustration by Kathryn Born, MA

Arterial walls are very strong, thick, and very elastic to withstand the great pressure to which the arteries are subjected from the blood pushing through them.

Although larger in diameter, veins have thinner walls and are less elastic. Since the push of the blood from the heart's contraction is gone by the time it enters the veins, many of them have valves to prevent backflow. While a pressure gradient still exists (lower in the heart) it's not enough to keep the blood flowing in that direction — especially against gravity. Muscle contractions from everyday movements squeeze on the veins, pushing blood forward and then they relax, pulls blood into the (temporarily) empty space — much like how an eyedropper works.

Withstanding the pressure

You've likely had your *blood pressure* taken before — the strap wrapped around your upper arm then inflated just to the point that you consider ripping it off from the pain. The purpose of this contraption is to block blood flow to your forearm. Then, as air is let out, blood begins to flow turbulently creating the audible *sounds of Kortokoff* (that's why the stethoscope is positioned on your antecubital region, or inner elbow). When the pressure around your arm matches the *systolic blood pressure*, the maximum force that the blood puts on the walls of the arteries due to ventricular contraction, some blood gets through, hitting the walls which you can hear. Then, when the external pressure matches the *diastolic blood pressure*, the force applied to the walls between contractions, the sounds stop because the blood is now allowed to flow smoothly. So your blood pressure is recorded as the systolic pressure over the diastolic, or 120/80 mmHg on average.

REMEMBER

The importance of blood pressure is not only to keep the blood circulating; it is the driving force behind capillary exchange, which is discussed in the next section.

Blood pressure that is too high *(hypertension)* damages the artery walls leading to a cascade of problems that can ultimately lead to heart failure. Unfortunately for us, the body doesn't seem to care much about this issue, as plenty of resources are being exchanged at the capillaries. There is no built-in mechanism for homeostasis of blood pressure if it is chronically high. On the other hand, low blood pressure *(hypotension)* means the body's tissues aren't receiving enough oxygen and other nutrients — a problem that must be solved post haste. A series of hormones are released via the *renin-angiotensin system*, which leads to *vasoconstriction* (decreasing the diameter of the arteries) and water retention in the kidneys, both of which lead to an increase in blood pressure.

Capillary exchange

You may have found yourself wondering. "Just how exactly do the tissues get the resources if the blood never leaves the vessels?" Well first, you must understand that blood is a tissue composed of red blood cells, white blood cells, platelets, and plasma. Some of the plasma, which is mostly water, can exit leaving the rest of the tissue behind. This occurs easily in the capillaries because their entire wall is just one layer of simple squamous epithelium. Though the cells are held tightly together, there is still room for the plasma to escape between them — especially with the force of blood pressure. This process, called *filtration*, allows molecules in the blood to be pushed out into the *interstitial fluid* of the tissues. Cells can then bring in those necessary materials.

In addition to filtration, osmosis and diffusion play a vital role in capillary exchange. Oxygen diffuses along its concentration gradient into the interstitial fluid while carbon dioxide diffuses into the blood. This also moves other nutrients out and wastes in. Due to the outflow of water in the first half of the capillary, the water of the interstitial fluid is drawn back in, along with anything dissolved in it.

Now is your chance to practice circulating through the circulatory system:

---

48 Number the structures in the correct sequence of blood flow from the heart to the radial artery for pulse. Start at the heart with the aortic semilunar valve.

_____ Axillary artery

_____ Subclavian artery

_____ Ascending aorta

_____ Brachial artery

_____ Aortic arch

49 Number the structures in the correct sequence of blood flow from the great saphenous vein back to the heart.

_____ External iliac vein

_____ Right atrium

_____ Common iliac vein

_____ Femoral vein

_____ Inferior vena cava

50 What are the sounds of Kortokoff?

a. the pulsation of the blood on the walls of the veins during normal blood flow

b. the sound of turbulent blood hitting the arteries when blood flow is restricted

c. the sounds of contraction of the atria and ventricles

d. the lub-dub sounds of the heart's valves closing

51 True or False: The renin-angiotensin system is activated in order to decrease blood pressure when it begins damaging arterial walls.

52 Which process does NOT play a major role in capillary exchange?

a. diffusion

b. osmosis

c. filtration

d. endocytosis

e. all of the above play a role

Answers to Questions on the Cardiovascular System

The following are answers to the practice questions presented in this chapter.

1. The system for gaseous exchange in the lungs: **b. Pulmonary circuit**

2. The system for maintaining a constant internal environment in other tissues: **c. Systemic circuit**

3. The membranous sac that surrounds the heart: **a. Pericardium**

4. The wall that divides the heart into two cavities: **e. Septum**

5. The layer providing the heart's wall with nutrients: **d. Epicardium**

6. What's the difference between the pulmonary and systemic circuits? **e. The pulmonary circuit travels to and from the lungs, and the systemic circuit keeps pressure and flow up for the body's tissues.** The Latin *pulmon*– means "lung." None of the other answers even mentions the lungs.

7. True or False: The heart is centrally located in the chest. **False.** Actually, two-thirds of the heart lies to the left of the body's midline.

8. Which blood vessel contains oxygenated blood: **b. pulmonary vein.** It's the only vein in the body that does so.

9. Only one of the heart wall's three layers contains muscle tissue. Which one? **c. Myocardium.** *Myo*– means "muscle," *epi*– is "upon," *endo*– is "within," and *peri*– is "around."

10-25. Following is how Figure 11-2, the heart, should be labeled.

10. c. Brachiocephalic trunk; 11. j. Superior vena cava; 12. d. Right pulmonary veins; 13. p. Right atrium; 14. g. Right coronary artery; 15. n. Inferior vena cava; 16. k. Left common carotid artery; 17. f. Left subclavian artery; 18. h. Aortic arch; 19. l. Left pulmonary arteries; 20. o. Pulmonary trunk; 21. a. Left pulmonary veins; 22. m. Left atrium; 23. i. Great (left) cardiac vein; 24. b. Left ventricle; 25. e. Right ventricle

26-29. Following is how Figure 11-3, the heart valves, should be labeled.

26. b. Pulmonary valve; 27. c. Aortic valve; 28. a. Tricuspid valve; 29. d. Mitral valve

30. Why is the myocardium of the ventricles thicker than the myocardium of the atria? **c. The myocardium is the muscular tissue of the heart, and more force is needed in chambers pumping blood greater distances.** The correct answer is the only one that correctly identifies myocardium as muscle tissue. Remember: *myo*– means "muscle."

31. The superior vena cava enters the heart by way of the **d. right atrium.**

32. What holds the cusps of the atrioventricular valves in place? **b. The chordae tendineae.** While the papillary muscles aid in the process, it's the chordae tendineae that keep them from flapping up into the atrium.

(33) What covers the atrioventricular opening between the right atrium and the right ventricle? **b. tricuspid valve.** Sorry if this seemed to be a trick question, but even if you have trouble remembering the heart's right openings from its left ones, you simply need to remember that the bicuspid and the mitral valve are the same thing, so "tricuspid valve" is the only correct answer here.

(34) Returns blood to the heart from the head, thorax, and upper extremities: **c. Superior vena cava**

(35) Valve located between the right atrium and right ventricle: **a. Tricuspid valve**

(36) Valve located between the right ventricle and pulmonary trunk: **d. Pulmonary valve**

(37) Returns blood to the heart from the trunk and lower extremities: **e. Inferior vena cava**

(38) Valve located between the left atrium and left ventricle: **b. Mitral valve**

(39) Semilunar valve located in the left ventricle: **f. Aortic valve**

(40) Returns blood from the lungs: **g. Pulmonary vein**

(41) Why is the sinoatrial node (SA node) called the heart's "pacemaker"? **b. It initiates an electrical impulse that causes atrial walls to contract simultaneously.** The other answers address rhythm, but the correct answer shows how the S-A node coordinates the heart's contractions.

(42) True or False: The chordae tendineae open the AV valves. **False.** While it looks that way since they're attached to the underside of the valve, they do not pull them open. They keep them closed; more specifically, they keep the flaps of the AV valves from being pulled up into the atrium.

(43) Choose the correct conductive pattern. **c. S-A node → A-V node → bundle of His → bundle branches → Purkinje fibers.**

(44) Which valve action results from the pressure in the ventricle being lower than that of the atria: **c. AV valves open**

(45) Draw brackets on the EKG in Figure 11-5 to show atrial and ventricular systole. **The bracket for atrial systole should begin at the peak of the first, smaller wave and stop at the peak of the second, larger one. The bracket for ventricular systole should begin near where the second wave reaches its peak and end where the third wave returns to baseline.**

REMEMBER

The waves on an EKG do not directly show contraction. The first two show the depolarization of the fibers (spread of impulse) that triggers the contractions. The third wave shows the repolarization of the ventricles.

(46) True or False: The atrial syncytium is comprised of the conducting fibers of the atrium that ensure the atria contract before the ventricles. **False.** While the first part of the statement is true, the purpose is to make both atria contract at the exact same time. It's the AV node that ensures the delay.

47 Starting with a drop of blood returning to the heart from the legs, trace its path until its eventual return to the legs. **Inferior vena cava → right atrium → tricuspid valve → right ventricle → pulmonary valve → pulmonary trunk → pulmonary arteries → lungs → pulmonary veins → left atrium → mitral valve → left ventricle → aortic valve → aorta → body**

REMEMBER

To be fair, the odds of that particular drop of blood making it back to the same place in the legs on its next trip are pretty slim. But the important point here is that it must pass through the heart twice to get sent back out to the body.

48 Number the structures in the correct sequence of blood flow from the heart to the radial artery for pulse. Start at the heart with the aortic semilunar valve. **1. Ascending aorta; 2. Aortic arch; 3. Subclavian artery; 4. Axillary artery; 5. Brachial artery.**

49 Number the structures in the correct sequence of blood flow from the great saphenous vein back to the heart. **1. Femoral vein; 2. External iliac vein; 3. Common iliac vein; 4. Inferior vena cava; 5. Right atrium.**

50 What are the sounds of Kortokoff: **b. the sound of turbulent blood hitting the arteries when blood flow is restricted**

51 True or False: The renin-angiotensin system is activated in order to decrease blood pressure when it begins damaging arterial walls: **False**. This is utilized to increase blood pressure.

52 Which process does NOT play a major role in capillary exchange: **d. endocytosis**. Endocytosis is the active transport mechanism cells use to bring in large molecules. Cells of the tissues would likely utilize this to bring in some of the nutrients the blood carried. It's not used to get them out of the blood; that would just put them in the cells of the capillary wall.

Chapter **12**

Keeping Up Your Defenses: The Lymphatic System

You see it every rainy day — water, water everywhere, rushing along gutters and down storm drains into a complex underground system that most would rather not give a second thought. Well, it's time to give hidden drainage systems a second thought: Your body has one, called *the lymphatic system*. This chapter explains what you need to know about this system.

Duct, Duct, Lymph

You already know that the body is made up mostly of fluid. *Interstitial* or *extracellular fluid* moves in and around the body's tissues and cells constantly. The story of the lymphatic system (shown in Figure 12-1) begins deep within the body's tissues at the farthest reaches of blood capillaries, where plasma leaks out of blood capillaries at the rate of nearly 51 pints a day, carrying various substances to and away from the smallest nooks and crannies. About 90 percent of that fluid gets reabsorbed into blood capillaries taking CO_2 and other waste molecules with it. But one or two liters of extra fluid remains in the tissues as interstitial fluid. However, in order for the body to maintain a sufficient volume of water within the cardiovascular system, eventually this interstitial fluid must get back into the blood. Additionally, the fluids' contents need to be managed for the homeostasis of our internal environment. That's where the lymphatic system steps in, acting as a recycling system to drain, transport, cleanse, and return this fluid to the bloodstream.

REMEMBER

The lymphatic system is more than a drainage network. It's a body-wide filter that traps and destroys invading microorganisms as part of the body's immune response network. It can remove pathogens from the body, help absorb and digest excess fats, and maintain a stable blood volume despite varying environmental stresses. Without it, the cardiovascular system would grind to a halt.

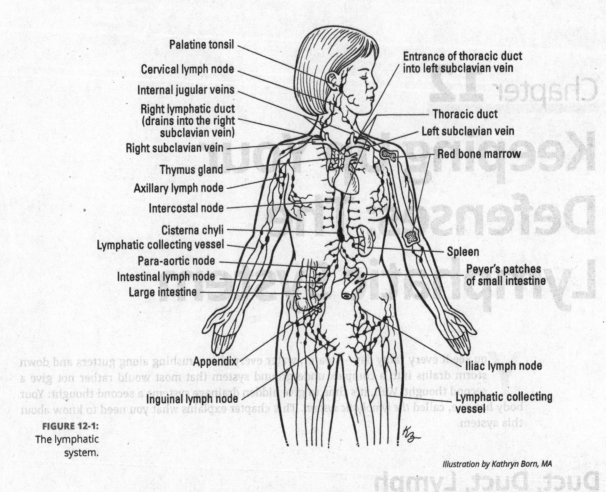

Palatine tonsil
Cervical lymph node
Internal jugular veins
Right lymphatic duct (drains into the right subclavian vein)
Right subclavian vein
Thymus gland
Axillary lymph node
Intercostal node
Cisterna chyli
Lymphatic collecting vessel
Para-aortic node
Intestinal lymph node
Large intestine

Entrance of thoracic duct into left subclavian vein
Thoracic duct
Left subclavian vein
Red bone marrow
Spleen
Peyer's patches of small intestine

Appendix
Iliac lymph node
Lymphatic collecting vessel
Inguinal lymph node

FIGURE 12-1:
The lymphatic system.

Illustration by Kathryn Born, MA

To collect the fluid, minute vessels called *lymph capillaries* are woven throughout the body, with a few caveats and exceptions. There are no lymph capillaries in the teeth, outermost layer of the skin, certain types of cartilage, any other avascular tissue, or bones. And because bone marrow makes lymphocytes, which we explain in the next section, it's considered part of the lymphatic system. Plus, *lacteals* (lymphatic capillaries found in the villi of the intestines) absorb fats to mix with lymph, forming a milky fluid called *chyle*. (See Chapter 14 for details on lacteals.)

WARNING

It was long thought that the CNS (central nervous system) did not contain any lymphatic vessels. However, in 2017, researchers were finally able to use advances in imaging technology to confirm that there are, in fact, lymphatic vessels in and around the brain.

Unlike blood capillaries, lymph capillaries dead-end (terminate) within tissue. Made up of a single layer of loosely overlapping endothelial cells anchored by fine filaments, lymph capillaries behave as if their walls are made of cellular, one-way valves. Endothelium provides a smooth, slick, friction-free surface in the capillaries. When the pressure outside a capillary is greater than it is inside, the filaments anchoring the cells allow them to open, permitting interstitial fluid to seep in. Increasing pressure inside the capillary walls eventually forces the entry gates to close. Once in the capillaries, the trapped fluid is known as *lymph*, and it moves into larger, vein-like *lymphatic vessels*. The lymph moves slowly and without any kind of central pump through a combination of peristalsis, the action of semilunar valves, and the squeezing influence of surrounding skeletal muscles, much like that which occurs in veins.

TIP

Plasma, interstitial fluid, and lymph are all basically the same thing — body fluid made mostly of water carrying various molecules. The difference is location. If the fluid is in the blood, then we call it plasma. Out in the tissues? Now it's interstitial fluid. Located in a lymphatic vessel? Presto! Now it's lymph.

Lymphatic capillaries are found in conjunction with blood capillaries and the vessels form networks around blood vessels. The larger lymphatic vessels contain *lymph nodes*, which are bean-shaped filters (more on these in the next section). The vessels then merge into *lymphatic trunks*, which merge into either the *thoracic duct* or the *right lymphatic duct*. The trunks are named for the region they serve (lumbar, intestinal, intercostal, subclavian, jugular, bronchomediastinal). Both ducts drain the lymph into the respective subclavian veins (thoracic goes into the left subclavian), where it becomes plasma again.

To see how much of this information is seeping in, answer the following questions:

1. The lymphatic system plays an important role in regulating

   a. intracellular energy function.

   b. interstitial fluid protein.

   c. metabolizing unused fats from the small intestine.

   d. intercellular transportation of oxygen.

2. Under what circumstances do the walls of lymphatic capillaries open?

   a. When the filaments holding the cells together begin to dehydrate

   b. When the pressure outside the capillaries is greater than inside

   c. When the capillaries sense extra salts in the bloodstream

   d. When reticular fibers begin to force their way between the cells

3 How does lymph move around inside its vessels?

    a. Rapidly, through the action of tiny pumps called lymph nodes

    b. Slowly, and only as additional lymph enters the system

    c. Slowly, through peristalsis, the action of semilunar valves, and the squeezing of surrounding skeletal muscles

    d. Steadily and rhythmically, mirroring the heart rate

4 What is the largest lymphatic vessel in the body?

    a. Right lymphatic duct

    b. Spleen

    c. Thoracic duct

    d. Lymph node

    e. Bone marrow

5 The lymphatic system serves several crucial functions, but what does it NOT normally do?

    a. Return interstitial fluid to the blood

    b. Destroy bacteria

    c. Remove old erythrocytes

    d. Filter the lymph

    e. Produce erythrocytes

Poking at the Nodes

Lymph nodes (see Figure 12-2) are the site of filtration in the lymphatic system. Also sometimes incorrectly referred to as *lymph glands* — they don't secrete anything, so technically they're not glands — these bean-shaped sacs are surrounded by connective tissue (and therefore are tough to spot). Lymph nodes contain macrophages, which destroy bacteria, cancer cells, and other matter in the lymph fluid. *Lymphocytes* (a type of white blood cell), which produce an immune response to microorganisms, also are found in lymph nodes. The indented part of each node, called the *hilum*, is where the efferent vessels exit and where the blood vessels (that supply the node tissue) enter and exit. Afferent vessels bring the lymph in on the convex side. The *stroma* (body) of each node is surrounded by a fibrous capsule that dips into the node to form *trabeculae*, or *septa* (thin dividing walls) that divide the node into compartments. *Reticular* (netlike) fibers are attached to the trabeculae and form a framework for the lymphoid tissue and clusters of macrophages and B lymphocytes called *lymphatic nodules*.

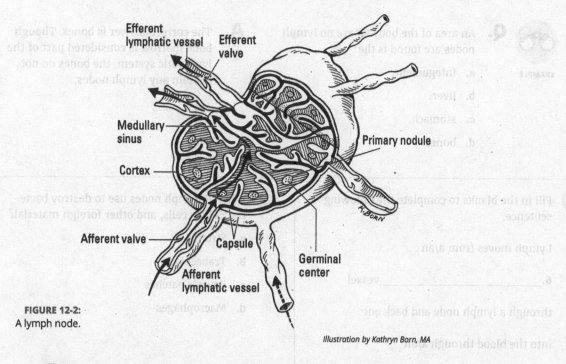

FIGURE 12-2:
A lymph node.

Illustration by Kathryn Born, MA

TIP

If you have trouble remembering your afferent from your efferent, think of the "a" as standing for "access" and the "e" as standing for "exit."

Although some lymph nodes are isolated from others, most nodes occur in groups, or clusters, particularly in the inguinal (groin), axillary (armpit), and mammary gland areas. (You can see some lymph nodes in Figure 12-1.) The following are the primary lymph node regions:

>> **Cervical:** Found in the neck, filter lymph from the head

>> **Axillary:** Found in armpits, filter lymph from arms and mammary region

>> **Supratrochlear:** Found above inner elbow, filter lymph from hands

>> **Inguinal:** Found in inguinal region, filter lymph from lower limbs and external genitalia

>> **Pelvic:** Found in pelvic cavity, filter lymph from pelvic organs

>> **Abdominal:** Found in abdominal cavity, filter lymph from abdominal organs

>> **Thoracic:** Found in mediastinum, filter lymph from heart and lungs

Each node acts like a filter bag filled with a network of thin, perforated sheets of tissue — a bit like cheesecloth — through which lymph must pass before moving on. White blood cells line the sheets of tissue, including several types that play critical roles in the body's immune defenses. This filtering action explains why, when infection first starts, lymph nodes often swell with the cellular activity of the immune system launching into battle with the invading microorganisms.

Think you have a node-tion (sorry!) about what's happening here? Test your knowledge:

EXAMPLE

Q. An area of the body where no lymph nodes are found is the

a. integument.

b. liver.

c. stomach.

d. bones.

A. The correct answer is bones. Though bone marrow is considered part of the lymphatic system, the bones do not contain any lymph nodes.

6-7 Fill in the blanks to complete the following sentence.

Lymph moves from a/an

6._____ vessel

through a lymph node and back out

into the blood through a/an

7._____ vessel.

8 What do lymph nodes use to destroy bacteria, cancer cells, and other foreign material?

a. Stroma

b. Trabeculae

c. Peyer's patches

d. Macrophages

9 Why do lymph nodes sometimes swell when an infection is present?

a. To make room for additional lymph fluid

b. Because of increased viscera that results from gastric activity

c. To accommodate cellular activity of the immune-system-attacking microorganisms

d. Because white blood cells are larger than red blood cells

10 Lymph nodes have two primary functions. What are they?

a. To conserve white blood cells and to produce lymphocytes

b. To produce bilirubin and to cleanse lymph fluid

c. To produce lymphocytes and antibodies and to filter lymph fluid

d. To conserve iron and to remove erythrocytes

e. To produce erythrocytes and to conserve iron

11 The two lymphatic ducts empty into the

a. lymph nodes

b. spleen

c. thymus

d. right atrium

e. subclavian veins

12 The germinal center of lymph nodules produces what type of cells?

a. Monocytes

b. Lymphocytes

c. Phagocytes

d. Trabeculae

e. Erythrocytes

13 The connective tissue fiber that forms the framework of the lymphoid tissue is

    a. cartilaginous.

    b. collagenous.

    c. reticular.

    d. elastic.

14 Which of the following regions of lymph nodes filter lymph from the lower limbs?

    a. axillary

    b. pelvic

    c. cervical

    d. inguinal

    e. abdominal

Having a Spleen-did Time with the Lymphatic Organs

Although the lymph nodes are the most numerous lymphatic organs, several other vital organs exist in the lymphatic system, including the spleen, thymus gland, and tonsils.

Reuse and recycle: The spleen

The spleen, the largest lymphatic organ in the body, is a 5-inch, roughly egg-shaped structure to the left of and slightly behind the stomach. Like lymph nodes, it has a hilum through which the splenic artery, splenic vein, and efferent vessels pass. (The efferent vessels transport lymph away from the spleen to lymph nodes; as we note in the preceding section, remember "e" for "exit.") Also like lymph nodes, the spleen is surrounded by a fibrous capsule that folds inward to section it off. Arterioles leading into each section are surrounded by masses of developing lymphocytes that give those areas of so-called *white pulp* their appearance. On the outer edges of each compartment, tissue called *red pulp* consists of blood-filled cavities. Unlike lymph nodes, the spleen doesn't have any afferent (access) lymph vessels, which means that it doesn't filter lymph, only blood.

Blood flows slowly through the spleen to allow it to remove microorganisms, exhausted erythrocytes (red blood cells), and any foreign material that may be in the stream. Among its various functions, the spleen can be a blood reservoir. When blood circulation drops while the body is at rest, the spleen's vessels can dilate to store any excess volume. Later, during exercise or if oxygen concentrations in the blood begin to drop, the spleen's blood vessels constrict and push any stored blood back into circulation.

REMEMBER

The spleen's primary role is as a biological recycling unit, capturing and breaking down defective and aged red blood cells to reuse their components. Iron stored by the spleen's macrophages goes to the bone marrow, where it's turned into hemoglobin in new blood cells. By the same token, bilirubin for the liver is generated during breakdown of hemoglobin. The spleen produces red blood cells during embryonic development but shuts down that process after birth; in cases of severe anemia, the spleen sometimes starts up production of red blood cells again.

Fortunately, the spleen isn't considered a vital organ; if it is damaged or has to be surgically removed, the liver and bone marrow can pick up where the spleen leaves off.

T cell central: The thymus gland

Tucked just behind the breastbone and between the lungs in the upper chest (the *superior mediastinum*, if you want to be technical), the thymus gland was a medical mystery until recent decades. Its two oblong lobes are largest at puberty when they weigh around 40 grams (somewhat less than an adult mouse). Through a process called *involution*, however, the gland atrophies and shrinks to roughly 6 grams by the time an adult is 65. (You can remember that term as the inverse of evolution.)

The thymus gland serves its most critical role — as a nursery for immature T lymphocytes, or T cells — during fetal development and the first few years of a human's life. Prior to birth, fetal bone marrow produces *lymphoblasts* (early stage lymphocytes) that migrate to the thymus. Shortly after birth and continuing until adolescence, the thymus secretes several hormones, collectively called *thymosin*, that prompt the early cells to mature into full-grown T cells that are *immunocompetent*, ready to go forth and conquer invading microorganisms. (These hormones are the reason the thymus is considered part of the endocrine system, too; check out Chapter 10 for details on this system.)

As with other lymphatic structures, the thymus is surrounded by a fibrous capsule that dips inside to create chambers called *lobules*. Within each lobule is a cortex made of T cells held in place by reticular fibers and a central medulla of unusually onion-like layered epithelial cells called *thymic corpuscles*, or *Hassall's corpuscles*, as well as scattered lymphocytes.

Open wide and move along: The tonsils and Peyer's patches

Like the thymus gland, the tonsils, which are misunderstood masses of lymphoid tissue, are largest around puberty and tend to atrophy as an adult ages. Unlike the thymus, however, the tonsils don't secrete hormones but do produce lymphocytes and antibodies to protect against microorganisms that are inhaled or eaten. Although only two are visible on either side of the pharynx, there are actually five tonsils:

>> The two you can identify, which are called *palatine tonsils*

>> The *adenoid* or *pharyngeal tonsil* in the posterior wall of the nasopharynx

>> The *lingual tonsils,* which are round masses of lymphatic tissue arranged in two approximately symmetrical collections that cover the posterior one-third of the tongue

Invaginations (ridges) in the tonsils form pockets called *crypts*, which trap bacteria and other foreign matter.

Peyer's patches are masses of lymph nodules just below the surface of the ileum, the last section of the small intestine. When harmful microorganisms get into the intestine, Peyer's patches can mobilize an army of B cells and macrophages to fight off infection.

You've absorbed a lot in this section. See how much of it is getting caught in your filters:

---

15 What role does the spleen have besides that of a biological recycling unit?

a. It undergoes involution as a person ages to make way for enlarged lymph nodes.

b. It divides the pharynx from the tonsils to ensure that lymph fluid moves in the right direction.

c. It produces neutrophils to maintain immunocompetent T cells.

d. It has vessels that dilate to store excess blood volume when the body is at rest.

16 Why does the thymus stop growing during adolescence and atrophy with age?

a. It serves its most critical role during fetal development and the first few years of life.

b. The increasing number of lymph nodes over time takes over its role in the body.

c. The reticular fibers surrounding it contract and crush it over time.

d. It slowly runs out of Hassall's corpuscles.

---

17 Lymphoid tissue(s) located in the pharynx that protect(s) against inhaled or ingested pathogens and foreign substances is/are called the

a. thymus.

b. tonsils.

c. Peyer's patches.

d. spleen.

18 Lymphatic nodules found in the ileum of the small intestines are

a. Peyer's patches.

b. lymph nodes.

c. thymus.

d. macrophages.

---

19 What is the significance of the spleen having no afferent vessels?

a. Foreign materials can't get inside and cause problems.

b. It means that the spleen filters only blood, not lymph.

c. It's why humans can survive removal of the spleen.

d. The spleen evolved that way to prevent phagocytosis.

20 The lymphatic organ found in the superior mediastinum is the

a. tonsil.

b. spleen.

c. thymus.

d. reticular formation.

 21 Why is the spleen's white pulp that color?

a. Blood has drained from the red pulp back into the bloodstream.

b. Cartilage is beginning to form at the borders.

c. Lymphocytes clump there and give it its color.

d. The elastic connective tissue has stretched and relaxed repeatedly over the course of a person's lifetime.

 22 The lymphatic organ(s) responsible for removal of aged and defective red blood cells from the bloodstream is/are the

a. tonsils.

b. spleen.

c. Peyer's patches.

d. lymph nodes.

23 Why does the thymus secrete hormones?

a. To make T lymphocytes immunocompetent

b. To help the tonsils make more T lymphocytes

c. To support the Peyer's patches as they produce B lymphocytes

d. To signal the spleen that it's time to release more red blood cells

Investigating Immunity

The lymphatic system is the battleground for our immune processes. But nearly every tissue in our bodies have some role to play (which is why you no longer see the immune system as one of the body's listed systems). We still have much to learn about all the mechanisms our bodies use to defend us from invaders, but this section provides a brief overview of what we do know.

First, it's important to understand that our body's defenses are divided into two categories: innate and adaptive.

Innate defenses

Innate, or *non-specific*, *defenses* are the tools our bodies use to attack foreign invaders regardless of their ilk. Germs can be bacteria, viruses, fungi, or other microorganisms and other foreign particles (pollen, toxins) can be problematic as well. Our innate defenses target all of these.

First and foremost is our skin — the body's largest organ and our first line of defense. Along with our other mechanical barriers, such as mucus and tears, most of the potential invaders are never even allowed entry. Should one make it into the body we have other innate strategies for our second line of defense:

>> **Chemical barriers**

- Enzymes (in saliva, gastric juice) break down cell walls.

- *Interferons* block replication (especially of virus and tumor cells).

- *Defensins* poke large holes in cell membranes.

- *Collectins* group together pathogens for easier phagocytosis.

>> **Inflammation:** Dilates blood vessels, sending more resources to the area where the pathogen was identified

>> **Fever:** Weakens microorganisms and stimulates phagocytosis

>> **Natural killer cells (NKs):** Secrete *perforins* to poke tiny holes in, or perforate, cell membranes

>> **Phagocytosis:** Consumption of foreign invaders by specialized white blood cells

Unfortunately, the occasional pathogen makes it past these defenses so our bodies mount a targeted attack. Furthermore, if we relied solely on our innate defenses, there would be massive amounts of collateral damage to our own cells (which is responsible for many of our symptoms of illness in the first place).

Adaptive defenses

Our *adaptive*, or *specific*, *defense* mounts a two-tiered attack targeting the specific pathogen. This minimizes collateral damage but takes time to get started. This process is dependent on molecules that stick off the surface of cells called *antigens*. All cells have them, unique to their variety, and that's how our immune cells distinguish self versus non-self. A type of white blood cell called a *macrophage* destroys a pathogen by phagocytosis; however, it leaves the antigens intact and displays them on itself. This way, it's one of our own cells that looks foreign searching for the matching lymphocytes to initiate our adaptive response.

There are two varieties of lymphocyte that carry out this response: *T cells* which mature in the thymus and *B cells* which mature in the bone marrow (see the connection?). The action of T cells is called *cell-mediated immunity* and of B cells it is called *humoral immunity*.

Cell-mediated

Once a macrophage finds a T cell with receptors that match its displayed antigens, they bind together. The lymphocyte, called a *helper T cell*, releases a chemical called *interleukin*-2, which activates another matching T cell. This stimulates the now *cytotoxic T cell* to begin proliferating (making copies of itself). These cytotoxic Ts (sometimes called killer Ts) will bind with antigens on the invader and release perforins, killing the pathogen. So only cells with this particular antigen will be targeted. When the battle has waned, *suppressor T cells* signal the adaptive immune process to stop. Some T cells will remain as *memory T cells* once the pathogen has been defeated. This way, if it invades again, it won't take long for the macrophage to find a match and the pathogen will be destroyed before you even show any symptoms — thus providing you *immunity*.

Humoral

B cells, with matching receptors, will bind to the pathogen or the antigen-presenting macrophage. When the helper T cell is activated it also releases *cytokines* which, in turn, activate the B cell. It begins to proliferate into *plasma B cells* and *memory B cells*. The memory Bs hang around with the memory T cells in the lymph nodes for protection later. The plasma Bs begin manufacturing *antibodies*, which are proteins that will bind to the antigens on the pathogens. When bound with antibodies, the pathogen is now *neutralized*. Since they have two binding sites, antibodies can also cause *agglutination*, clumping together the invaders for more efficient phagocytosis. They also can activate the *complement cascade*, a series of chemical reactions that can directly destroy the pathogen.

The faster we can locate the matching B and T cells, the less damage the pathogen can cause. Lymphocytes are generated with random receptor shapes and researchers argue that we all have one cell in us somewhere to match any pathogen we could possibly encounter — the issue is, can we find it before the pathogen does irreversible damage.

Hopefully these questions won't seem too foreign:

24 Why are macrophages considered the bridge between our innate and adaptive immune responses?

_____

_____

_____

_____

25-30 Match the immune strategy to its description.

a. block entry into body

b. disrupt replication of pathogen

c. gather pathogens together for destruction

d. tear large holes in cell membrane

e. released by NKs to poke tiny holes in cell membranes

f. white blood cell engulfs the entire pathogen

25 _____ collectins

26 _____ defensins

27 _____ interfeons

28 _____ mucus

29 _____ perforins

30 _____ phagocytosis

31 One of the mechanisms used by antibodies is to clump together pathogens for more efficient phagocytosis. This is called:

a. complement cascade.

b. neutralization.

c. precipitation.

d. agglutination.

e. collectins.

32-36 Fill in the blanks to complete the following paragraph.

Pathogens, like our own cells, are covered in 32._____ which enable our white blood cells to distinguish between self and non-self. When activated, 33._____ that have the matching receptors release perforins to destroy the specific pathogen. Meanwhile, B cells synthesize 34._____ that will target the pathogen as well. Both types of lymphocyte keep 35._____ around for a faster response on the next exposure, granting you 36._____ from that particular pathogen.

Answers to Questions on the Lymphatic System

The following are answers to the practice questions presented in this chapter.

(1) The lymphatic system plays an important role in regulating **b. interstitial fluid protein.** By keeping the interstitial fluid volume between tissue cells in balance, the lymphatic system also keeps the body in homeostasis.

(2) Under what circumstances do the walls of lymphatic capillaries open? **b. When the pressure outside the capillaries is greater than inside.** The difference in pressure causes the filaments anchoring the thin layer of cells to open. Increasing pressure inside the capillary walls eventually forces them closed again.

(3) How does lymph move around inside its vessels? **c. Slowly, through peristalsis, the action of semilunar valves, and the squeezing of surrounding skeletal muscles.** You already know the first answer is wrong because lymph nodes aren't pumps, and because you know the lymphatic system doesn't have pumps, you know the last answer is wrong because there is nothing steady or rhythmic about it. You're left with one of the "slowly" answers. The second answer can't be right because the system wouldn't have room to take in more lymph if what was already there wasn't moving through some kind of action. Voilà! You have the right answer.

(4) What is the largest lymphatic vessel in the body? **c. Thoracic duct.** Yes, this duct is the largest lymphatic vessel. "Spleen" isn't the correct answer because that's the largest lymphatic *organ*.

(5) The lymphatic system serves several crucial functions, but what does it *not* normally do? **e. Produce erythrocytes.** Those are red blood cells, which develop in the bone marrow.

(6-7) Lymph moves from a/an **6. afferent lymphatic** vessel through a lymph node and back out into the blood through a/an **7. efferent lymphatic** vessel. Not sure which vessel is which? A comes first in the alphabet, and you can think of it as the lymph's Access point. E comes next, and it provides an Exit point.

(8) What do lymph nodes use to destroy bacteria, cancer cells, and other foreign material? **d. Macrophages.** You can arrive at the correct answer through a process of elimination. Neither stroma nor trabeculae can be right because you already know that they're both part of connective tissue. Peyer's patches are limited to the small intestine. So macrophages must be the front-line warriors here.

(9) Why do lymph nodes sometimes swell when an infection is present? **c. To accommodate cellular activity of the immune-system-attacking microorganisms.** You can eliminate a couple of answers right off the bat here. The first answer can't be right; why would there be a need for additional lymph fluid? You also know the word "gastric" in the second answer relates to digestion, not the lymphatic system.

(10) Lymph nodes have two primary functions. What are they? **c. To produce lymphocytes and antibodies and to filter lymph fluid.** Be careful that both parts of the answer are correct.

(11) The two lymphatic ducts empty into the **e. subclavian veins.**

12 The germinal center of lymph nodules produces what type of cells? **b. Lymphocytes.** Your answer hint is in the question: *Lymph* nodules produce *lymph*ocytes.

13 The connective tissue fiber that forms the framework of the lymphoid tissue is **c. reticular.** It provides both a tissue framework and a type of netting to hold clusters of lymphocytes.

14 Which of the following regions of lymph nodes filter lymph from the lower limbs? **d. inguinal.**

15 What role does the spleen have besides that of a biological recycling unit? **d. It has vessels that dilate to store excess blood volume when the body is at rest.** Enlarged lymph nodes don't travel, so you know the first answer is wrong. Plus, you know the spleen is in the abdomen, not the throat, so toss out that second answer. Wavering between the third and fourth answers? Just remember all those TV shows where internal bleeding is chalked up to a damaged spleen, and you've got it!

16 Why does the thymus stop growing during adolescence and atrophy with age? **a. It serves its most critical role during fetal development and the first few years of life.**

Here's a memory tool that only word-play students will love: "The thymus runs out of thyme."

17 Lymphoid tissue(s) located in the pharynx that protect(s) against inhaled or ingested pathogens and foreign substances is/are called the **b. tonsils.** When you remember that the pharynx is the throat, this question becomes more obvious.

18 Lymphatic nodules found in the ileum of the small intestines are **a. Peyer's patches.** It's almost like they're "patched" onto the ileum.

19 What is the significance of the spleen having no afferent vessels? **b. It means that the spleen filters only blood, not lymph.** No afferent vessels means no access for lymph to get in there.

20 The lymphatic organ found in the superior mediastinum is the **c. thymus.**

Break this question into parts and it becomes easier to locate which gland is being referenced: *Superior* means "upper," *media–* means "middle" (or "midline"), and *–stinum* refers to the sternum, or breastbone.

21 Why is the spleen's white pulp that color? **c. Lymphocytes clump there and give it its color.** Red pulp is always red, so say goodbye to that first answer. The spleen doesn't need cartilage, so toss out the second answer. And the final answer can't be right because what would be the evolutionary point in scarring up the spleen as a person ages?

22 The lymphatic organ(s) responsible for removal of aged and defective red blood cells from the bloodstream is/are the **b. spleen.** It recycles critical components from the spent red blood cells and sends them to the bone marrow to be turned into fresh cells.

23 Why does the thymus secrete hormones? **a. To make T lymphocytes immunocompetent.** Plus, it's where these cells get the "T" in their name.

24 Why are macrophages considered the bridge between our innate and adaptive immune responses? **Macrophages are, themselves, an innate defense. They consume anything with foreign antigens so it's non-specific. However, it then displays those antigens trying to find a helper T with matching receptors to activate in order get the adaptive response going.**

(25) collectins **c. gather pathogens together for destruction**

(26) defensins **d. tear large holes in cell membranes**

(27) interferons **b. disrupt replication of pathogen**

(28) mucus **a. block entry into body**

(29) perforins **e. released by NKs to poke tiny holes in cell membranes**

(30) phagocytosis **f. white blood cell engulfs the entire pathogen**

(31) One of the mechanisms used by antibodies is to clump together pathogens for more efficient phagocytosis. This is called **d. agglutination.** Watch out for collectins; they serve the same purpose, but this is an innate chemical defense produced by white blood cells.

(32-36) Pathogens, like our own cells, are covered in **32. anitgens** which enable our white blood cells to distinguish between self and non-self. When activated, **33. Cytotoxic T cells** that have the matching receptors release perforins to destroy the specific pathogen. Meanwhile, B cells synthesize **34. antibodies** that will target the pathogen as well. Both types of lymphocyte keep **35. memory cells** around for a faster response on the next exposure, granting you **36. immunity** from that particular pathogen.

Chapter **13**

Oxygenating the Machine: The Respiratory System

P eople need lots of things to survive, but the most urgent need from moment to moment is oxygen. Without a continual supply of this vital element, we don't last long. But if we have reserves of the other things we need — carbohydrates, fats, and proteins — why don't we have some kind of storehouse of oxygen, too? Simple. It's readily available in the air around us, so we've never needed to evolve a means for storing it. Nonetheless, our stored food supplies would be useless without oxygen; our bodies can't metabolize the energy they need from these substances without a constant stream of oxygen to keep things percolating along.

All that metabolizing creates another equally important need, however. We must have a means for getting rid of our bodies' key gaseous waste — carbon dioxide, or CO_2. If it builds up in our systems, we die. It must be removed from our bodies almost as fast as it's formed. Conveniently, breathing in fulfills our need for oxygen and breathing out fulfills our need to expel carbon dioxide.

In this chapter, you get a quick review of Mother Nature's dual-purpose system and plenty of opportunities to test your knowledge about the lungs and other parts of the respiratory system.

Breathing In the Basics

Respiration, or the exchange of gases between an organism and its environment, occurs in three distinct processes:

>> **Breathing:** The technical term is *pulmonary ventilation*, or the movement of air into and out of the lungs. Breathing is comprised of two distinct actions: *inspiration* and *expiration*.

>> **Exchanging gases:** This takes place between the lungs, the blood, and the body's cells in two ways:

- **Pulmonary, or external, respiration:** The exchange in the lungs when blood gains oxygen and loses carbon dioxide

- **Systemic, or internal, respiration:** The exchange that takes place in and out of capillaries when the blood releases some of its oxygen and collects carbon dioxide from the tissues

>> **Cellular respiration:** Oxygen is used in the catabolism of substances like glucose for the production of energy (see Chapter 2), creating CO_2 as a byproduct.

A single respiratory cycle consists of one inhalation followed by an expiration. The regular, restful breathing rate is controlled by the pons, while the medulla oblongata will signal any necessary changes to that pattern. To complete a normal inhalation, the *diaphragm* (the broad skeletal muscle that forms the bottom of the thoracic cavity) is triggered to contract. This pushes down on the contents of the abdominal cavity, thus increasing the volume of the lungs. An increase in volume causes a decrease in pressure (known as *Boyle's Law*). So, as a result of the diaphragm contracting, the pressure of the air already inside the lungs drops below that of atmospheric pressure (the pressure of the air outside our bodies). Because gasses will naturally diffuse to areas of lower pressure, air flows into the lungs. Thus we do not need to "suck in" air with each normal breath. For a deeper inhalation, we contract the intercostal muscles (between the ribs), which pull the ribcage out, further increasing the volume and dropping the pressure inside the lungs even lower so more air can come in. Once inside the lungs, gasses can be exchanged between the air and the blood.

Exhalation is a passive process; that is, we don't tell the lungs to breathe out. We simply stop telling the diaphragm (and the intercostal muscles if they're engaged) to contract. When they relax, volume decreases and pressure increases. Further, the lungs contain a great deal of elastic tissue. As the muscles relax, the elastic tissue snaps back. This *elastic recoil* briefly drops the pressure inside the lungs to below atmospheric pressure and air flows out.

Because we can consciously amplify our ventilation, there are differing, measurable lung volumes. These are summarized below:

>> **Tidal volume:** The amount of air moved in or out during normal ventilation

>> **Inspiratory reserve:** The additional amount of air that can be inhaled with a forceful inspiration

>> **Expiratory reserve:** The additional amount of air that can be exhaled with a forceful expiration

» **Residual volume:** The amount of air that must remain in the lungs at all times (the lungs never deflate with exhalation, even if it's forceful)

» **Vital capacity:** The maximum amount of air that can be moved into (or out of) the lungs

» **Total lung capacity:** The total amount of air the lungs can hold; equal to the vital capacity plus the residual volume

Air enters the lungs through the *trachea* which branches to the right and left, forming the *primary bronchi*. These then branch several times forming the *bronchial tree* which distributes the air throughout the lungs:

Primary bronchi → Secondary bronchi → Tertiary bronchi → Bronchioles →
Alveolar ducts → Alveoli

The alveoli are the air sacs and are the site of gas exchange. The bulk of the respiratory tract is wrapped in rings of hyaline cartilage (to maintain their shape) and is lined with ciliated pseudostratified tissue. The lining contains numerous *goblet cells* to produce mucus to trap unwanted particles in the air (like dust). The cilia then sweep the mucus up and out to be expelled — sometimes with the aid of a cough. The further down in the bronchial tree you proceed, the fewer the goblet cells you see. (We definitely don't want mucus in our air sacs.) The bronchioles lack the supportive cartilage rings and are lined with simple cuboidal, while the alveoli are comprised of simple squamous tissue.

The lungs themselves are divided into lobes: The right has three while the left has two (to leave room for the heart). Between the right and left lungs is the *mediastinum*, the central compartment of the thoracic cavity that houses the heart, major blood and lymphatic vessels, nerves, trachea, and esophagus. Each lung is covered in the *visceral pleura*. Since the lungs occupy most of the remaining space inside the thoracic cavity, there is little to no space between the visceral pleura and the *parietal pleura*, which lines the thoracic cavity (refer to Chapter 1 to brush up on these layers and compartments). The attraction between the two pleural membranes facilitates lung expansion during inspiration.

Now take a deep breath and see if you can answer some questions:

EXAMPLE

Q. The air that moves in and out of the lungs during normal, quiet breathing is called

   a. tidal volume.

   b. inspiratory reserve.

   c. vital capacity.

   d. lung capacity.

   e. residual volume.

A. The correct answer is tidal volume. The question asks only about air moved during normal, quiet breathing, not the kind of forceful air movement involved in measuring lung capacity. Think of the normal ebb and flow of the ocean's tide as opposed to the waves of a raging storm.

1 What does the mediastinum have to do with the lungs?

    a. It's the bone that protects them from collapsing during blunt trauma.

    b. It's the region between the right and left lungs.

    c. It's the control mechanism that reduces minimal air when the body is stressed.

    d. It's the membrane surrounding each lung.

    e. It's the muscle structure that moves the diaphragm to inflate and deflate them.

2 What is the purpose of mucus in the respiratory tract?

    a. to reduce friction of the air passing through

    b. to coat the alveoli to facilitate gas exchange

    c. to trap gas molecules for transport into the blood stream

    d. to trap dust and pollutants in the air for removal

3 Which term can be best described as the total amount of air than can be moved into or out of the lungs?

    a. tidal volume

    b. inspiratory reserve

    c. residual volume

    d. vital capacity

    e. total lung capacity

4 True or False: The lungs inflate with inspiration and deflate with expiration.

5 In which part of the respiratory system is cartilage never found?

   a. bronchioles

   b. primary bronchi

   c. secondary bronchi

   d. tertiary bronchi

   e. trachea

6-10 Fill in the blanks to complete the following sentences:

The trachea divides into two

6._____, which then divide

into **7.**_____ with a branch

going to each lobe of the lung. Upon entering

the lobe, each divides into

8._____, subdividing into

smaller tubes called **9.**_____.

They terminate in an elongated sac called

the atrium surrounded by

10._____ or air sacs.

11 If a pin were to pierce the body from the outside in the thoracic region, which three structures — in order — would it pass through?

   a. Lung, pleural cavity, terminal bronchioles

   b. Visceral pleura, lung, pleural cavity

   c. Parietal pleura, pleural cavity, visceral pleura

   d. Pleural cavity, trachea, lungs

   e. Primary bronchi, secondary bronchi, terminal bronchioles

12 Explain how normal inspiration is an active process while expiration is passive.

_____

_____

_____

_____

Preparing the Air

Before entering the trachea, air must first pass through the nose, pharynx, and larynx which help to prepare and filter the air. Though we are able to breathe through our mouths, that's more of a back-up plan so here we focus on the nose.

Knowing about the nose (and sinuses)

You may care a great deal about how your nose is shaped, but the shape actually makes little difference to your body. The nose is simply the most visible part of your respiratory tract. Beyond those oh-so-familiar nostrils — which are formally called *external nares* — the *septum* divides the nasal cavity into two chambers. Inside the nostril is a slight dilation extending to the apex of the nose called the *vestibule*; it's lined by ciliated pseudostratified epithelium covered with hairs, plus mucous and sebaceous glands that help trap dust and particles before they can enter the lungs. The cilia sweep the mucus into the pharynx, where it is swallowed. Just under the lining are numerous blood vessels which serve to warm the air and the mucus moisturizes it.

Each nasal cavity is divided into an *olfactory region* and a *respiratory region*.

>> **Olfactory region:** The olfactory region lies in the upper part of the nasal cavity. Fine filaments distributed over its mucous membrane are actually olfactory receptors devoted to the sense of smell. The bipolar olfactory cells' axons thread through openings in the *cribriform plate* (from the Latin *cribrum* for "sieve") of the ethmoid bone and then come together in the *olfactory bulb*, which acts as an odorant classification filter (for perception of smells). These send signals to the higher olfactory centers of the brain's cerebral cortex for further discrimination and enhancement of the olfactory process. Impulses flow from the olfactory bulb through the olfactory tract to two main destinations:

- Via the *thalamus* to the *olfactory cortex* in the temporal lobe, where smells are consciously detected

- Via the sub-cortical route to the *hypothalamus* and to other regions of the limbic systems where emotional aspects of the smells are analyzed

>> **Respiratory region:** The remainder of the nasal cavity comprises the respiratory region.

REMEMBER

The respiratory region of the nasal cavity performs several important functions:

>> It drains mucous secretions from the sinuses.

>> It drains lacrimal secretions from the eyes.

>> It prepares inhaled air for the lungs by warming, moistening, and filtering it. Dust and bacteria are caught in the mucus and passed outward from the nasal cavity by the motion of the cilia. Some of that gunk is taken up by *lymphatic tissue* in the nasal cavity and respiratory tubes for delivery to the lymph nodes, which destroy invading germs.

Ah, the sinuses. They can be such headaches. Lined with a pseudostratified ciliated columnar epithelium, *paranasal sinuses* are air-filled cavities in the bone that reduce the skull's weight and act as resonators for the voice. Each of the four groups of sinuses is named for the bone containing it (see Chapter 7), as follows:

>> **Frontal sinuses** are located in the frontal bone behind the eyebrows. (If you've ever flown with a sinus infection in an airplane, these are the suckers that hit you right behind the eyes.)

>> **Maxillary sinuses** are located in the *maxillae,* or upper jawbone.

>> **Ethmoid** and **sphenoid sinuses** are located in the ethmoid and sphenoid bones in the cranial cavity's floor.

Beyond the sinuses and connected to them are *nasal ducts* that extend from the medial angle of the eyes to the *nasal cavity*. These ducts let serous fluid from the eyes' *lacrimal glands* (which you know as tears) flow into the nasal cavity. Beyond the nasal cavity is the *nasopharynx*, which connects — you guessed it — the nasal cavity to the *pharynx*.

With a bit of a refresher on the nasal and sinus passages, do you think you can hit the following practice questions on the nose?

---

13 Which of the following statements about the mucous membranes of the nasal cavity is NOT true?

a. They contain an abundant blood supply.

b. They moisten the air that flows over them.

c. They're composed of stratified squamous epithelium.

d. They filter the air of particulates.

14 Why does the nose have sebaceous glands?

a. Their secretions smooth air flow into the nostrils.

b. They enhance the sense of smell.

c. Their secretions form a protective layer to warm, moisten, and help filter air.

d. They improve resonation during speech.

15-29 Use the terms that follow to identify the structures of the respiratory tract shown in Figure 13–1.

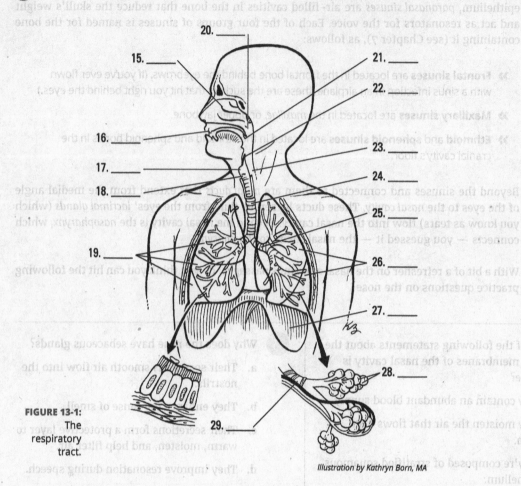

20. _____

15. _____

21. _____

22. _____

16. _____

23. _____

17. _____

24. _____

18. _____

25. _____

19. _____

26. _____

27. _____

28. _____

29. _____

FIGURE 13-1:
The respiratory tract.

Illustration by Kathryn Born, MA

a. Esophagus

b. Larynx

c. Nasal cavity

d. Oropharynx

e. Diaphragm

f. Right lung

g. Epiglottis

h. Mouth

i. Alveoli

j. Nasopharynx

k. Thyroid cartilage

l. Left lung

m. Trachea

n. Bronchioles

o. Left primary bronchus

Dealing with throaty matters

In laymen's terms, it's the throat. But you know better, right? The "throat" consists of these key parts:

>> **Pharynx:** The pharynx is an oval, fibromuscular tube about 5 inches long and tapering to ½ inch in diameter at its *anteroposterior end*, which is a fancy biology term meaning "front to back." In fact, the point where the pharynx connects to the esophagus is the narrowest part of the entire digestive tract. The pharynx consists of three main divisions:

  • The *nasopharynx*, where the nasal cavity opens into the pharynx

  • The *oropharynx*, where the oral cavity opens into the pharynx

  • The *laryngopharynx*, which begins at the epiglottis and leads down to the esophagus

  Eustachian tubes connected to the middle ears enter the nasopharynx on each side. (And when the tubes become blocked by an infection, sudden pressure changes can be extremely painful.) On the back wall of the nasopharynx is a mass of lymphoid tissue called the *pharyngeal tonsil*, or *adenoids*.

>> **Larynx:** Connecting the pharynx with the trachea, this collection of nine cartilages is what makes a man's prominent Adam's apple. Also called the voice box, the larynx looks like a triangular box flattened dorsally and at the sides that becomes narrow and cylindrical toward the base. Ligaments connect the cartilages to several muscles, allowing their movements to be controlled. The inside of the larynx is lined with a mucous membrane that continues into the trachea.

Three of the larynx's nine cartilages go solo — the *thyroid*, the *cricoid*, and the *epiglottis* — while the others come in pairs — the *arytenoids*, the *corniculates*, and the *cuneiforms*.

>> **Thyroid cartilage:** The thyroid cartilage (*thyroid* in Greek means "shield-shaped") is largest and consists of two plates called *laminae* that are fused just beneath the skin to form a shield-shaped process, the Adam's apple. Immediately above the Adam's apple, the laminae are separated by a V-shaped notch called the *superior thyroid notch*.

>> **Cricoid cartilage:** The ring-shaped cricoid cartilage is smaller but thicker and stronger, with shallow notches at the top of its broad back that connect, or articulate, with the base of the arytenoid cartilages.

>> **Arytenoid cartilages:** The arytenoid cartilages are both shaped like pyramids, with the vocal folds attached at the back and the controlling arytenoid muscles that move the arytenoids attached at the sides, moving the vocal cords.

>> **Corniculate cartilages:** On top of the arytenoids are the corniculate cartilages, small conical structures for attachment of muscles regulating the tension of the vocal cords.

>> **Cuneiform cartilages:** Nestled in front of these and inside the *aryepiglottic* fold, the cuneiform cartilages stiffen the soft tissues in the vicinity.

>> **Epiglottis cartilage:** The epiglottis, sometimes called the lid on the voice box, is formed from elastic cartilage in the shape of a leaf. It is attached at its stem end and projects upward behind the root of the tongue. It opens during respiration and reflexively closes during swallowing to keep food and liquids from getting into the respiratory tract. Its elastic cartilage gives it a flexible attachment to the anterior and superior borders of the *hyoid bone* and larynx.

Two types of folds play different roles inside the larynx.

>> The *true vocal folds*, or cords, are V-shaped when relaxed. When talking, the folds stretch for high sounds or slacken for low sounds, causing the glottis — the opening in the larynx — to form an oval. Sounds are produced when air is forced over the folds, causing them to vibrate.

>> Just above these folds are the *vestibular folds*, also known as the false vocal cords, that don't produce sounds. Muscle fibers within these folds help close the glottis during swallowing — the last chance for preventing food from entering the trachea.

Following are some practice questions dealing with the throat:

---

 30 What happens when the arytenoids move?

   a. The Adam's apple moves.

   b. The cricoid cartilage flexes.

   c. The vocal cords move.

   d. The epiglottis closes.

   e. The cuneiform cartilages stiffen.

 31-35 Match the anatomical structure with its function.

   a. Voice box lid

   b. Respiratory center

   c. Prevent collapse of trachea

   d. Fight germs

   e. Adam's apple

31 _____ Epiglottis

32 _____ Adenoids

33 _____ Medulla oblongata

34 _____ C-shaped cartilaginous rings

35 _____ Thyroid cartilage

36 The _____ is the opening between the two vocal folds, and the _____ covers that opening as needed.

a. epiglottis, glottis

b. bronchi, epiglottis

c. alveoli, pharynx

d. glottis, pharynx

e. glottis, epiglottis

37-49 Use the terms that follow to identify the structures of the larynx shown in Figure 13-2. *Note:* Some terms may be used more than once.

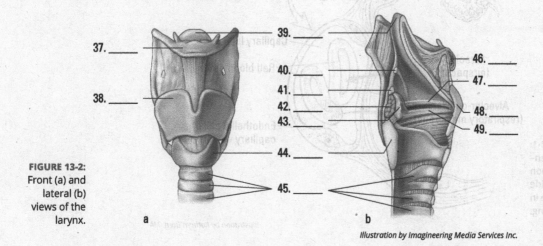

FIGURE 13-2: Front (a) and lateral (b) views of the larynx.

a

b

Illustration by Imagineering Media Services Inc.

a. Thyroid cartilage

b. Cricoid cartilage

c. Hyoid bone

d. Epiglottis

e. Arytenoid cartilage

f. Vestibular fold (false vocal cord)

g. Corniculate cartilage

h. Cuneiform cartilage

i. Arytenoid muscle

j. True vocal cord

k. C-shaped tracheal cartilages

Swapping the bad for the good

Now that we know how the air is prepared, and enters and leaves the lungs, it's time to look at what happens while it's there. The final destination of the air is the *alveoli* (singular: *alveolus*). At the end of the terminal bronchioles are the *alveolar ducts* which route air into the *alveolar sac*. Linked to this are the alveoli forming what looks like a cluster of grapes — with each alveolus being a single grape. Wrapped around each cluster of alveoli is a network of capillaries — this is the site of gas exchange which you can see illustrated in Figure 13-3. The numerous alveoli create a large surface area for this.

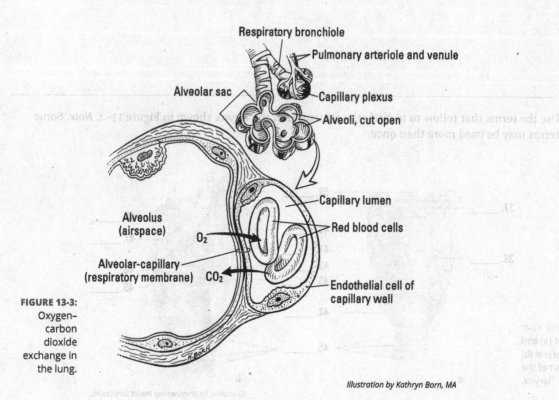

Respiratory bronchiole

Pulmonary arteriole and venule

Alveolar sac

Capillary plexus

Alveoli, cut open

Capillary lumen

Red blood cells

Alveolus (airspace)

O₂

Alveolar-capillary (respiratory membrane)

CO₂

Endothelial cell of capillary wall

FIGURE 13-3: Oxygen–carbon dioxide exchange in the lung.

Illustration by Kathryn Born, MA

Like the walls of the capillaries (see Chapter 11), the walls of the alveoli are simple squamous. This means the *respiratory membrane* is merely two layers of thin, flattened cells thick. With only two layers of cells between the air and the blood, gas exchange occurs easily due to the differences in partial pressure of O_2 and CO_2 (which you can think of as a concentration gradient). Because the partial pressure of CO_2 (pCO_2) is lower in the air than in the blood, the molecules diffuse through the respiratory membrane into the air in the alveoli to be exhaled. Since the partial pressure of O_2 (pO_2) is higher in the air, it diffuses readily into the blood where it is picked up by hemoglobin in the red blood cells to be carried to the tissues.

REMEMBER Blood comes to the lungs through two sources: the *pulmonary arteries* and the *bronchial arteries*. The pulmonary trunk comes from the right ventricle of the heart and then branches into the two pulmonary arteries carrying deoxygenated blood (the only arteries that contain blood loaded with carbon dioxide) from various parts of the body to the lungs (see Chapter 11 for an introduction to the cardiovascular system). That blood goes through pulmonary arterioles and then through pulmonary capillaries for gas exchange. After that, oxygenated blood leaves the lung through pulmonary venules, returning to the left atrium through the *pulmonary veins* (the only veins that contain oxygenated blood), completing the cycle. Bronchial arteries branch off the thoracic aorta of the heart, supplying the lung tissue with nutrients and oxygen.

See whether you're carrying away enough information about respiration by tackling the following questions:

 50 What is hemoglobin's primary role?

a. Filtration

b. Gaseous transport

c. Ventilation

d. Asphyxia

51-54 Fill in the blanks to complete the following sentences:

Upon inhalation, molecules of

51._____ diffuse into the

lung's tissues. From there, these molecules

then diffuse into 52._____

cells, which contain a pigment called

53._____. Simultaneously,

a second substance formed during cellular

respiration, 54._____, is

released into the lungs to be expelled during

exhalation.

55-61 Use the terms that follow to identify the structures of the bronchiole shown in Figure 13-4.

55. _____

56. _____

57. _____

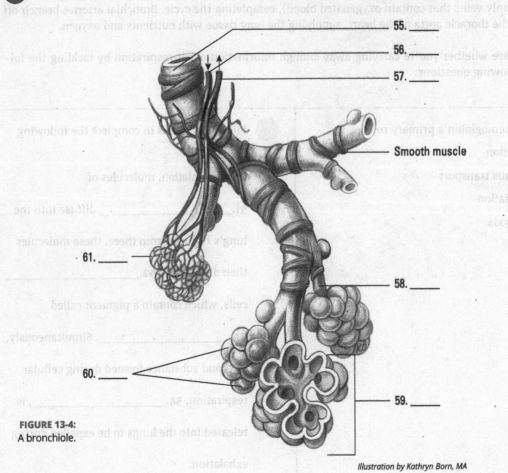

— Smooth muscle

61. _____

58. _____

60. _____

59. _____

FIGURE 13-4:
A bronchiole.

Illustration by Kathryn Born, MA

a. Pulmonary venule

b. Alveolar sac

c. Tertiary bronchiole

d. Pulmonary capillary

e. Pulmonary arteriole

f. Alveolar duct

g. Alveoli

---

62 Blood in the pulmonary artery has a pO_2 of 40mmHg and a pCO_2 of 46mmHg. The air in the alveoli has a pO_2 of 105mmHg and a pCO_2 of 40mmHg. Which statement best describes what will occur?

a. Both O_2 and CO_2 will exit the alveoli.

b. Both O_2 and CO_2 will enter the alveoli.

c. O_2 will enter the alveoli and CO_2 will enter the blood.

d. O_2 will exit the alveoli and CO_2 will exit the blood.

Answers to Questions on the Respiratory System

The following are answers to the practice questions presented in this chapter.

1. What does the mediastinum have to do with the lungs? **b. It's the region between the right and left lungs.** It encases everything contained in the thoracic cavity except for the lungs.

2. What is the purpose of mucus in the respiratory tract? **d. to trap dust and pollutants in the air for removal**

3. Which term can be best described as the total amount of air than can be moved into or out of the lungs? **d. vital capacity.** The total lung capacity includes the residual volume which cannot ever exit the lungs.

4. True or False: The lungs inflate with inspiration and deflate with expiration. **False.** If all the air exited a lung, it would collapse — and so would the alveoli, meaning no gas exchange can occur.

5. In which part of the respiratory system is cartilage never found? **a. Bronchioles.** They're so small that they need to be more elastic than cartilage would allow them to be.

6-10. The trachea divides into two **6. primary bronchi,** which then divide into **7. secondary bronchi** with a branch going to each lobe of the lung. Upon entering the lobe, each divides into **8. tertiary bronchi,** subdividing into smaller tubes called **9. bronchioles.** They terminate in an elongated sac called the atrium surrounded by **10. alveoli,** or air sacs.

11. If a pin were to pierce the body from the outside in the thoracic region, which three structures — in order — would it pass through? **c. Parietal pleura, pleural cavity, visceral pleura.** Note that the question asks you to choose from the lists provided, not from the entire structure of the body.

12. Explain how normal inspiration is an active process while expiration is passive. **Breathing in requires muscle contraction which requires energy, hence active. Even without conscious control, the diaphragm contracts decreasing pressure inside of the lungs. From here, no more energy is expended. Air flows into the lungs because of the pressure differential. The medulla oblongata stops stimulating the diaphragm so it relaxes. That paired with the elastic recoil of the lung tissue increases pressure so air flows out; no contraction, no energy.**

13. Which of the following statements about the mucous membranes of the nasal cavity is not true? **c. They're composed of stratified squamous epithelium.** They are ciliated pseudostratified like most linings that secrete mucus from goblet cells.

14. Why does the nose have sebaceous glands? **c. Their secretions form a protective layer to warm, moisten, and help filter air.**

15-29. Following is how Figure 13-1, the respiratory tract, should be labeled.

15. c. Nasal cavity; 16. h. Mouth; 17. k. Thyroid cartilage; 18. b. Larynx; 19 f. Right lung; 20. j. Nasopharynx; 21. d. Oropharynx; 22. g. Epiglottis; 23. a. Esophagus; 24. m. Trachea; 25. o. Left primary bronchus; 26. l. Left lung; 27. e. Diaphragm; 28. i. Alveoli; 29 n. Bronchioles

(30) What happens when the arytenoids move? **c. The vocal cords move.** This cartilage is tough to spell and pronounce but easy to move.

(31) Epiglottis: **a. Voice box lid**

(32) Adenoids: **d. Fight germs**

(33) Medulla oblongata: **b. Respiratory center**

(34) C-shaped cartilaginous rings: **c. Prevent collapse of trachea**

(35) Thyroid cartilage: **e. Adam's apple**

(36) The **e. glottis** is the opening between the two vocal folds and the **e. epiglottis** covers that opening as needed. *Note:* The prefix *epi* means "on, upon, or above."

(37-49) Following is how Figure 13-2, the larynx, should be labeled.

37. **c. Hyoid bone**; 38. **a. Thyroid cartilage**; 39. **d. Epiglottis**; 40. **h. Cuneiform cartilage**; 41. **g. Corniculate cartilage**; 42. **e. Arytenoid cartilage**; 43. **i. Arytenoid muscle**; 44. **b. Cricoid cartilage**; 45. **k. C-shaped tracheal cartilages**; 46. **c. Hyoid bone**; 47. **f. Vestibular fold (false vocal cord)**; 48. **a. Thyroid cartilage**; 49. **j. True vocal cord**

(50) What is hemoglobin's primary role? **b. Gaseous transport.** Hemoglobin is the protein that bonds with and releases gasses within the bloodstream. That's how oxygen gets where it needs to go and how carbon dioxide gets shipped off to the lungs for waste removal.

TIP

Remember that the Latin root for blood is *hemo*; other than filtration, which occurs elsewhere, none of the other answer options apply to blood.

(51-54) Upon inhalation, molecules of **51. oxygen** diffuse into the lung's tissues. From there, these molecules then diffuse into **52. red blood** cells, which contain a pigment called **53. hemoglobin**. Simultaneously, a second substance formed during cellular respiration, **54. carbon dioxide,** is released into the lungs to be expelled during exhalation.

(55-61) Following is how Figure 13-4, the bronchiole, should be labeled.

55. **c. Tertiary bronchiole**; 56. **e. Pulmonary arteriole**; 57. **a. Pulmonary venule**; 58. **f. Alveolar duct**; 59. **b. Alveolar sac**; 60. **g. Alveoli**; 61. **d. Pulmonary capillary**

(62) Blood in the pulmonary artery has a pO_2 of 40mmHg and a pCO_2 of 46mmHg. The air in the alveoli has a pO_2 of 105mmHg and a pCO_2 of 40mmHg. Which statement best describes what will occur? **d. O_2 will exit the alveoli and CO_2 will exit the blood.** Gasses will always travel to where there is a lower partial pressure (which is just like the concentration gradient) if they're able to move through whatever barrier is in the way. Since the respiratory membrane is only two cells thick, oxygen and carbon dioxide diffuse through it easily, all without the use of ATP.

Chapter **14**

Fueling the Functions: The Digestive System

I t's time to feed your hunger for knowledge about how nutrients make their way into the human body. In this chapter, we help you swallow the basics about getting food into the system and digest the details about how nutrients move into blood flow. You also get plenty of practice following the nutritional trail from first bite to final elimination.

Digesting the Basics: It's Alimentary!

REMEMBER

Before jumping into a discussion of the alimentary canal, we need to review some basic terms:

» **Ingestion:** Taking in food

» **Mechanical Digestion:** Physically breaking down food into smaller pieces

» **Chemical Digestion:** Changing the composition of food — splitting large molecules into smaller ones — to make it usable by the cells

» **Deglutition:** Swallowing, or moving food from the mouth to the stomach

» **Absorption:** Occurs when digested food moves through the intestinal wall and into the blood

» **Egestion:** Eliminating waste materials or undigested foods at the lower end of the digestive tract; also known as *defecation*

The digestive tract, or alimentary canal, develops early on in a growing embryo. The primitive gut, or *archenteron*, develops from the *endoderm* (inner germinal layer) during the third week after conception, a stage during which the embryo is known as a *gastrula*. Whereas the respiratory tract (see Chapter 13) is a two-way street — oxygen flows in and carbon dioxide flows out — the digestive tract is designed to have a one-way flow (although when you're sick or your body detects something bad in the food you've eaten, what goes down sometimes comes back up). Under normal conditions, food moves through your body in the following order:

Mouth → Pharynx → Esophagus → Stomach → Small intestine → Large intestine

About 8 m long (~26 ft.), the alimentary canal is a muscular tube with a four-layered wall throughout. The outermost layer, the *serosa*, is a thin membrane with cells that secrete a clear, watery fluid to reduce friction as the organs rub against one another. It is also referred to as the *visceral peritoneum*. Proceeding inward, the next layer is the *muscularis*. It contains smooth muscle that contracts to push food through via *peristalsis* (see the section "Stomaching the Body's Fuel" later in this chapter). The outer part of this layer consists of *longitudinal muscle* that runs parallel to the length of the tube. Beneath that is a layer of *circular muscle* that wraps around. The stomach has a third layer of *oblique muscle* to aid in its mixing movements. The next layer is the *submucosa*. The submucosa is mostly areolar, or loose connective tissue to provide space for all of the nerves, glands, and vessels that reside there. The innermost layer is the *mucosa*, creating the continuous lining of the canal. It creates the *lumen*, or space within the tube, and its epithelial tissue is the site of absorption and secretion. In the esophagus, this layer is smooth but in the other organs it folds on itself to varying degrees. The creation of little pits and valleys increases surface area for interaction with the food occupying the lumen.

Chew on these sample questions about the alimentary canal:

 Match the alimentary canal terms with their descriptions.

a. Digestion

b. Ingestion

c. Deglutition

d. Absorption

e. Egestion

1 ___B___ Taking in food

2 ___E___ Elimination of waste

3 ___C___ Movement of food from mouth to stomach

4 ___D___ Means of transporting food into the blood

5 ___A___ Mechanical/chemical changing of food composition

As you read through the rest of this chapter, identify the parts of the digestive system. Referring to Figure 14-1, use the terms that follow to identify the corresponding parts of the digestive system.

a. Pancreas

b. Large intestine

c. Liver

d. Small intestine

e. Salivary glands

f. Gallbladder

g. Appendix

h. Anus

i. Esophagus

j. Rectum

k. Stomach

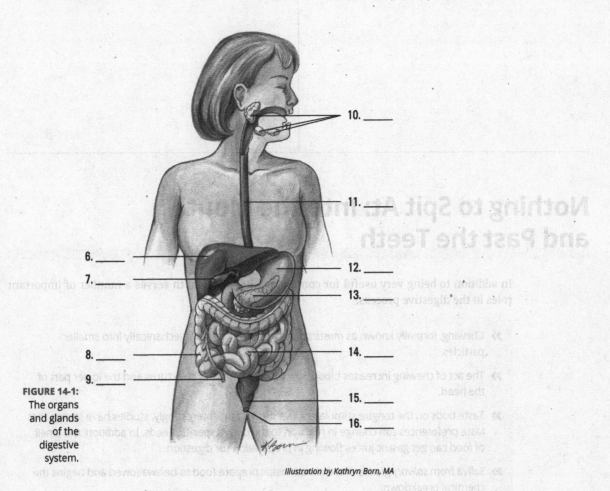

10. _____

11. _____

6. _____

7. _____

12. _____

13. _____

8. _____

14. _____

9. _____

15. _____

16. _____

FIGURE 14-1:
The organs
and glands
of the
digestive
system.

Illustration by Kathryn Born, MA

17 Identify the correct sequence of the movement of food through the body:

    a. Mouth → Pharynx → Esophagus → Stomach → Small intestine → Large intestine

    b. Mouth → Esophagus → Pharynx → Stomach → Small intestine → Large intestine

    c. Mouth → Pharynx → Esophagus → Stomach → Large intestine → Small intestine

    d. Mouth → Pharynx → Stomach → Esophagus → Small intestine → Large intestine

    e. Mouth → Esophagus → Stomach → Small Intestine → Pharynx → Large intestine

18 Indicate the order of the layered walls of the alimentary canal, from the innermost proceeding outward.

    _____ Muscularis, longitudinal

    _____ Muscularis, circular

    _____ Serosa

    _____ Mucosa

    _____ Lumen

    _____ Submucosa

Nothing to Spit At: Into the Mouth and Past the Teeth

In addition to being very useful for communicating, the mouth serves a number of important roles in the digestive process:

>> Chewing, formally known as *mastication,* breaks down food mechanically into smaller particles.

>> The act of chewing increases blood flow to all the mouth's structures and the lower part of the head.

>> Taste buds on the tongue stimulate saliva production. Interestingly, studies have shown that taste preferences can change in reaction to the body's specific needs. In addition, the smell of food can get gastric juices flowing in preparation for digestion.

>> Saliva from *salivary glands* in the mouth helps prepare food to be swallowed and begins the chemical breakdown.

The mouth's anatomy begins, of course, with the lips, which are covered by a thin, modified mucous membrane. That membrane is so thin that you can see the red blood in the underlying capillaries. (That's the unromantic reason for the lips' natural rosy glow.) As you see in the following sections, the mouth itself is divided into two regions defined by the *dental arches* of the upper and lower jaws. The *vestibule* is the region between these arches, and the cheeks and lips, whereas the *oral cavity* is the region inside the dental arches. (Figure 14-2 introduces the major structures of the mouth and the pharynx; flip to Chapter 13 for details on the pharynx.)

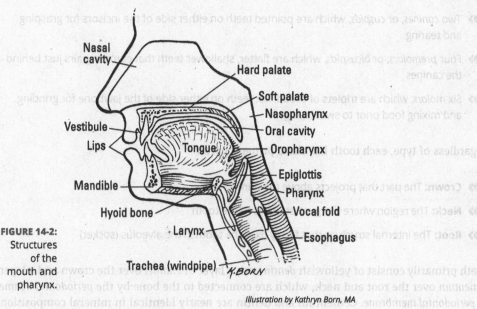

FIGURE 14-2: Structures of the mouth and pharynx.

Illustration by Kathryn Born, MA

Entering the vestibule

The inner surface of the lips is covered by a mucous membrane as well. Sickle-shaped pieces of tissue called *labial frenula* attach the lips to the gums. Within the mucous membrane are *labial glands,* which produce mucus to prevent friction between the lips and the teeth. The cheeks are made up of *buccinator muscles* and a *buccal pad,* a subcutaneous layer of fat, and the *masseters.* The buccinator muscles keep the food between the teeth during the act of chewing while the masseters are the main power behind it. Elastic tissue in the mucous membrane keeps the lining of the cheeks from forming folds that would be bitten during chewing (usually — most people have bitten the insides of their cheeks at one time or another).

The dental arches are formed by the *maxillae* (upper jaw) and the *mandible* (lower jaw) along with the *gingivae* (gums) and teeth of both jaws. The gingivae are dense, fibrous tissues attached to the teeth and the underlying jaw bones; they're covered by a mucous membrane extending from the lips and cheeks to form a collar around the neck of each tooth. The gums are very vascular (meaning that lots of blood vessels run through them) but poorly innervated (meaning that, fortunately, they're not generally very sensitive to pain).

Teeth rise from openings in the jawbone called *sockets,* or *alveoli.* You have a number of different kinds of teeth, and each has a specific contribution to the process of biting and chewing. Most humans get two sets of teeth in a lifetime. The first temporary, or *baby teeth,* set is known as *deciduous teeth.* Babies between 6 months and 2 years old "cut," or *erupt,* four incisors, two canines, and two molars in each jaw. These teeth are slowly replaced by permanent teeth from about 5 or 6 years of age until the final molars — referred to as *wisdom teeth* — erupt between 17 and 25 years of age.

An adult human has the following 16 teeth in each jaw (for a total set of 32 teeth):

>> Four *incisors*, which are chisel-shaped teeth at the front of the jaw for biting into and cutting food

>> Two *canines*, or *cuspids*, which are pointed teeth on either side of the incisors for grasping and tearing

>> Four *premolars*, or *bicuspids*, which are flatter, shallower teeth that come in pairs just behind the canines

>> Six *molars*, which are triplets of broad, flat teeth on either side of the jawbone for grinding and mixing food prior to swallowing

REMEMBER

Regardless of type, each tooth has three primary parts:

>> **Crown:** The part that projects above the gum

>> **Neck:** The region where the gum attaches to the tooth

>> **Root:** The internal structure that firmly fixes the tooth in the alveolus (socket)

Teeth primarily consist of yellowish *dentin* with a layer of *enamel* over the crown and a layer of *cementum* over the root and neck, which are connected to the bone by the *periodontal ligament*, or *periodontal membrane*. Cementum and dentin are nearly identical in mineral composition to bone; enamel consists of 94 percent calcium phosphate and calcium carbonate and is thickest over the chewing surface of the tooth.

Depending on the structure of the tooth, the root can be a single-, double-, or even triple-pointed structure. In addition, each tooth has a *pulp cavity* at the center that's filled with connective tissue, nerves, and blood vessels that enter the tooth through the root canal via an opening at the apex of the root of the tooth called the *apical foramen*. Now you know why it hurts so much when dentists have to drill down and take out that part of an infected tooth!

Moving along the oral cavity

The roof of the oral cavity is formed by both the *hard palate*, a bony structure covered by fibrous tissue and the ever-present mucous membrane, and the *soft palate*, a movable partition of fibromuscular tissue that prevents food and liquid from getting into the nasal cavity. (It's also the tissue that sometimes vibrates in sleep, causing a sonorous grating sound referred to as *snoring*.) The soft palate hangs at the back of the oral cavity in two curved folds that form the palatine arches. The *uvula*, a soft conical process (or piece of tissue), hangs in the center between those folds. (You can see the hard and soft palates in Figure 14-2.)

From the soft palate, the oral cavity arches back into the *oropharynx*. Here you'll find a mass of lymphatic tissue on each side — the *palatine tonsils*. That is, if a surgeon hasn't removed them because of frequent childhood infections. The following sections detail the remaining structures found in the oral cavity: the tongue and the salivary glands.

The tongue

As shown in Figure 14-2, the tongue, which is a tight bundle of interlaced muscles, and its associated mucous membrane, forms the floor of the oral cavity. Two distinct groups of muscles — *extrinsic* and *intrinsic* — are used in tandem for mastication (chewing), deglutition (swallowing), and to articulate speech.

>> **The extrinsic muscles:** Used to move the tongue in different directions, the extrinsic muscles originate outside the tongue and are attached to the mandible, styloid processes of the temporal bone and the hyoid. A fold of membrane called the *lingual frenulum* anchors the tongue.

>> **The intrinsic muscles:** A complex muscle network, the intrinsic muscles allow the tongue to change shape for talking, chewing, and swallowing.

The upper surface of the tongue contains numerous *papillae* (nipple-shaped protrusions) many of which contain taste buds. These contain *gustatory receptors* that relay chemical information to the cortex of the brain.

The salivary glands

The oral cavity has three pairs of salivary glands (which are exocrine glands) whose secretions combine to create *saliva*. Saliva, which contains water, electrolytes, mucus, and enzymes, begins the chemical digestion of the ingested food as well as softens it for ease in swallowing.

>> The *parotid* salivary glands are the largest salivary glands and are found in the posterior region of the mandible, just in front of and below each ear. They secrete a serous fluid through the parotid duct (which is opposite the second upper molar tooth) that makes up about 25 percent of the saliva that enters the oral cavity.

>> The smallest pair of the trio, the *sublingual* salivary glands, lies near the tongue under the oral cavity's mucous membrane floor. These produce about 5 percent of the saliva that enters the oral cavity.

>> The *submandibular* (or *submaxillary*) salivary glands are posterior to the sublingual, along the side of the lower jaw. They're about the size of walnuts and release fluid onto the floor of the mouth, under the tongue. They produce approximately 70 percent of the saliva that enters the oral cavity.

REMEMBER

And those secretions are nothing to spit at. Saliva does the following:

>> Dissolves and lubricates food to make it easier to swallow

>> Contains *amylase*, an enzyme that breaks down certain carbohydrates and *lipase*, which begins lipid (fat) breakdown

>> Contains *lysozymes* that break down the cell wall of some bacteria, killing them

>> Moistens and lubricates the mouth and lips, keeping them pliable and resilient for speech and chewing

>> Frees the mouth and teeth of food, foreign particles, and loose epithelial cells

>> Produces the sensation of thirst to prevent you from becoming dehydrated

Following are some practice questions regarding the vestibule and oral cavity:

EXAMPLE

Q. The function of the mouth is

   a. to mix solid foods with saliva.

   b. to break down milk protein via the enzyme rennin.

   c. To masticate or break down food into small particles.

   d. a and c

   e. a, b, and c

A. The correct answer is to mix solid foods with saliva and mastication (a and c). The mouth does lots of things, including mixing saliva into the food to add the enzyme amylase, but that's not rennin. With answer options like these, it's best to stick to the basics.

 Use the terms that follow to identify the parts of a tooth, as shown in Figure 14-3.

   a. Neck

   b. Dentin

   c. Crown

   d. Periodontal ligament

   e. Gingiva

   f. Enamel

   g. Root

   h. Apical foramen

   i. Pulp cavity

   j. Cementum

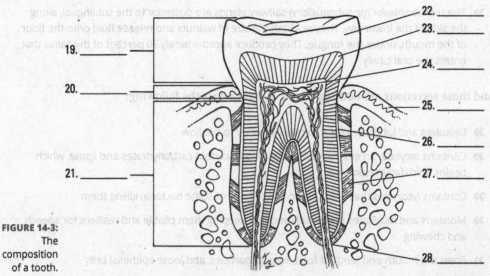

19. _____
20. _____
21. _____
22. _____
23. _____
24. _____
25. _____
26. _____
27. _____
28. _____

FIGURE 14-3:
The composition of a tooth.

Illustration by Kathryn Born, MA

29 What's the function of the soft palate, which vibrates when you snore?

a. It separates the uvula from the periodontal ligament.

b. It prevents premature swallowing of food before it's thoroughly masticated.

c. It forces saliva back over the tongue.

d. It prevents food and liquid from getting into the nasal cavity.

e. It secretes a watery fluid making the bolus easy to swallow.

30 What anchors the tongue?

a. The hard and soft palates

b. The intrinsic muscles and the sulcus terminalis

c. The extrinsic muscles and the lingual frenulum

d. The gingiva and the labial frenulum

e. The mandibular arches

31 Which of the following statements about teeth is NOT true?

a. The permanent teeth in each human jaw are four incisors, two canines, four premolars, and six molars.

b. Each tooth has a single cuspid anchoring it.

c. They are covered in both dentin and cementum.

d. The tooth cavity contains the tooth pulp.

e. The enamel consists of 94 percent calcium phosphate and calcium carbonate.

32 If it hasn't been surgically removed, where can the palatine tonsil be found?

a. In the posterior wall of the pharynx

b. In the smooth posterior third of the tongue

c. In the region between the rigid hard palate and the soft palate

d. In the region where the soft palate meets the oropharynx

e. In the region just superior to the uvula

33 Besides making it possible to spit, why do people have saliva?

   a. To facilitate swallowing

   b. To initiate the digestion of certain carbohydrates

   c. To moisten and lubricate the mouth and lips

   d. To combat bacteria

   e. All of the above

Stomaching the Body's Fuel

Deglutition, better known as swallowing, occurs in three phases:

REMEMBER **1.** **Food pushed into pharynx.** Following mastication, the tongue rolls the food into a *bolus*. It then pushes against hard palate propelling the food into the oropharynx. This is the only step in deglutition that is voluntary. Once the food reaches the oropharynx, the swallowing reflex occurs.

2. **Swallowing reflex.** *Tensor* and *levator* muscles contract causing the uvula and soft palate to raise, blocking off the *nasopharynx*. The hyoid bone and larynx elevate, forcing the *epiglottis* over the *trachea*. This step prevents food from entering the airway but momentarily stops breathing — hence the risk of choking if you gasp or someone makes you laugh. Next, the tongue pushes on the soft palate, sealing off the oral cavity. Now, the only place left for the food to go is into the *esophagus*. The longitudinal muscles of the pharynx contract, pulling the entry into the esophagus up to the bolus. (And you thought swallowing pushed the food down!) The *upper esophageal sphincter* relaxes and the swallowing reflex is complete.

3. **The *bolus* (food mass) heads "down the hatch."** The "hatch" is borrowed nautical slang for the esophagus. The esophagus performs a series of contractions called *peristalsis* to push the food down to the stomach. When the bolus enters the esophagus, it causes the tube to stretch. This triggers the *circular muscles* in the previous ring (the section of esophagus now just above the food) to contract and the next ring (the section now just below the food) to relax. This *receptive relaxation*, along with the contraction of the *longitudinal muscles* around the food, pushes the food down into the next segment and the process repeats until the bolus reaches the end of the esophagus. The *cardiac (lower esophageal) sphincter* is triggered to relax and the bolus enters the stomach.

After all that pushing and pulling, the food finally (after about 8 seconds, or just 1 if it's liquid) reaches the stomach, a pear-shaped bag of an organ that lies just beneath the ribs and diaphragm. When empty, the *mucosa* of the stomach lies in folds called *rugae*, which allow expansion of the stomach when you gorge and shrink when it's empty to decrease the surface area exposed to acid. Food enters the upper end of the stomach, called the *cardiac* region, through the cardiac sphincter, which generally remains closed to prevent gastric acids from moving up into the esophagus. (When gastric acids do move into places they don't belong, the painful sensation is referred to as "heartburn," which may help you remember the term "cardiac sphincter.") The dome-shaped area next to the cardiac region is called the *fundus*; it expands superiorly (upward) with really big meals. The middle part, or *body*, of the stomach forms a large curve on the left side called the *greater curvature*. The right, much shorter, border of the stomach's body is called the *lesser curvature*. The lower part of the stomach, shaped like the letter J, is the *pylorus*. The far end of the stomach remains closed off by the *pyloric sphincter* until its contents have been digested sufficiently to pass into the *duodenum* of the small intestine.

The secretions of the stomach are released by several glands embedded in the stomach's *epithelium* (the mucosa, or inner lining), which are named for the region in which they are found (cardiac, fundic, and pyloric). There are three main cell types found in these exocrine glands and their secretions combine to form *gastric juice*.

>> **Mucous cells:** Secrete mucus to protect the lining from the high acidity of the gastric juices. The main component of mucus is the protein *mucin*.

>> **Chief cells:** Secrete *pepsinogen,* a precursor to the enzyme pepsin that helps break down certain proteins into peptides. Chief cells also secrete lipase (to break fats down into lipids) and renin (to break down milk proteins), though production of the latter is often lost by adulthood.

>> **Parietal cells** secrete the hydrochloric acid, or HCl, that combines with pepsinogen to form pepsin to begin protein digestion. HCl also directly breaks down numerous other molecules in the food as well as kills many of the pathogens we inadvertently ingest.

The peristaltic contractions that get the bolus into the stomach aren't limited to the esophagus. Instead, peristalsis continues into the musculature of the stomach and stimulates the release of a hormone called *gastrin*. Within minutes, gastrin triggers secretion of gastric juices that reduce the bolus of food to a thick semiliquid mass called *chyme*, which passes through the pyloric sphincter into the small intestine within one to four hours of the food's consumption. While the main goal of the stomach is mechanical and chemical digestion, a small amount of absorption does occur here. Some water and certain salts find their way into our blood flow through the stomach's mucosa, as do alcohol and lipid-soluble drugs.

REMEMBER

Gastric juices are thin, colorless fluids with an extremely acidic pH that ranges from 1 to 4 (see Chapter 2 for details on pH). The quantity of acid released depends on the amount and type of food being digested.

34-41 Use the terms that follow to identify the anatomy of the stomach, as shown in Figure 14-4.

  a. Circular muscle layer

  b. Esophagus

  c. Rugae of the mucosa

  d. Cardiac (lower esophageal) sphincter

  e. Serosa

  f. Oblique muscle layer

  g. Pyloric sphincter

  h. Longitudinal muscle layer

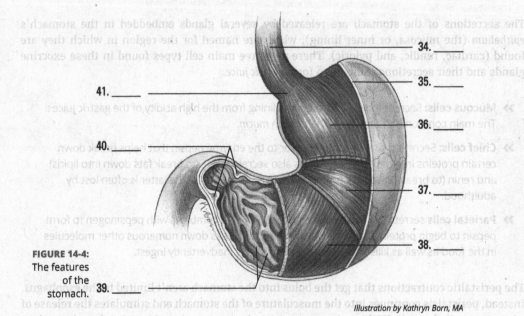

34. _____

35. _____

41. _____

36. _____

40. _____

37. _____

38. _____

FIGURE 14-4:
The features
of the
stomach.

39. _____

Illustration by Kathryn Born, MA

 42 What does peristalsis do to food?

  a. These contractions of the small intestine wring nutrients from partially digested food.

  b. This reflexive upward muscular motion mixes food with stomach acids to create vomit.

  c. This grinding action of cartilaginous tissue in the esophagus breaks down food before it enters the stomach.

  d. This sequential contraction of circular muscles helps move food through the esophagus.

  e. These contractions occur in the colon, hold feces there until defecation.

43 Which part of the digestive tract isn't doing its job when you have heartburn?

a. The esophagus

b. The pyloric sphincter

c. The peritoneal fold

d. The cardiac sphincter

e. The rugae

44 Where can chyme first be found along the digestive tract?

a. The pylorus

b. The peritoneum

c. The jejunum

d. The esophagus

e. The large intestine

Breaking Down the Work of Digestive Enzymes

So what exactly does all the work of digesting and breaking down food? That question brings you back into the realm of proteins (which we introduce in Chapter 2). Proteins that are *enzymes* act as catalysts, meaning that they initiate and accelerate chemical reactions without themselves being permanently changed in the reaction. Enzymes are very picky proteins indeed; they are effective only in their own pH range, they catalyze only a single chemical reaction, they act on a specific substance called a *substrate*, and they function best at 98.6 degrees Fahrenheit, which just happens to be normal body temperature.

The following sections take you on a tour of the organs that produce most of our digestive enzymes.

The small intestine

Most enzymatic reactions — in fact most digestion and practically all absorption of nutrients — takes place in the small intestine. Stretching 7 meters (which is nearly 23 feet!), this long organ extends from the stomach's pylorus to the *ileocecal junction* (the point where the small intestine meets the large intestine), gradually diminishing in diameter along the way.

REMEMBER

Three regions of the small intestine play unique roles as chyme moves through them:

>> **Duodenum:** The first section of the small intestine is also the shortest and widest section. As partially digested food enters the duodenum, its acidity stimulates the *enterogastric reflex* which feeds back to the stomach. The pyloric sphincter will tighten, preventing the stomach from transporting more chyme than the duodenum can handle. At this time, several hormones are released, though researchers are unsure of the exact role of all of them. Both the liver and pancreas share a common opening into the duodenum. Lined with large and numerous *villi* (fingerlike projections), the duodenum also has *Brunner's glands* that secrete a clear, alkaline mucus. The glands are most numerous near the entry to the stomach and decrease in number toward the opposite, or *jejunum,* end. Nearly all chemical digestion is complete by the time the chyme exits the duodenum.

>> **Jejunum:** This is where the bulk of the absorption of nutrients occurs. This region of the small intestine also contains villi, but unlike the duodenum, it has numerous large circular folds at the beginning that decrease in number toward the *ileum* end.

>> **Ileum:** Any nutrients not absorbed in the jejunum are absorbed in the ileum before the food passes into the large intestine. Another feature within the ileum are *Peyer's patches*, which are aggregates of lymph nodes that line this region, becoming largest and most numerous at the distal (far) end. They monitor the gut for pathogenic microorganisms to facilitate the mucosa's generation of immune responses. The ileum opens into the *cecum* of the large intestine through the *ileocecal sphincter*.

A microscopic look at the small intestine reveals circular folds called *plicae circularis*, which project 3 to 10 millimeters into the intestinal lumen, or cavity. These are permanent folds that don't smooth when the intestine is distended. Also present are villi that greatly increase the surface area through which the small intestine can absorb nutrients. Each villus contains a network of capillaries and a central lymph vessel, or *lacteal*, which absorbs fatty acids, to form a milk-white substance called *chyle.* The chyle is carried through lymphatic vessels to the thoracic duct, which empties into the subclavian vein. Simple sugars, amino acids, vitamins, minerals, and water are absorbed by the villi through the mucosa. The surface of the villus is simple columnar epithelium (if you can't recall what that means, flip to Chapter 5) and covered in *microvilli* to further increase surface area. Peristalsis continues into the small intestine, shortening and lengthening the villi to mix intestinal juices with food and increase absorption. Intestinal glands, called the *crypts of Lieberkühn,* lie in the depressions between villi. Packed inside these glands are antimicrobial *Paneth cells, goblet cells* to secrete mucus, and stem cells to facilitate replacement of the epithelial cells.

The extensive curves of the small intestine are held in place by the *mesentery* which also routes the blood and lymphatic vessels as well as nerves. Recent research has shown the mesentery is, in fact, an organ and may play a role in digestive disorders we have not yet pin-pointed a cause for. The *greater omentum* hangs down from the stomach, covering nearly all the intestines. It is laden with fat and contains numerous macrophages to battle pathogens. The greater omentum has been known to migrate to areas of damage, walling the area off to prevent the spread of infection. Both the mesentery and greater omentum are folded extensions of the peritoneum.

REMEMBER

Intestinal juices contain three types of enzymes:

>> **Enterokinase** has no enzyme action by itself, but when added to pancreatic juices, it combines with *trypsinogen* to form *trypsin,* which can break down proteins.

>> **Erepsins, or proteolytic enzymes,** don't directly digest proteins but instead complete protein digestion started elsewhere. They split polypeptide bonds, separating amino acids; an example is *peptidase.*

>> **Inverting enzymes** split disaccharides into monosaccharides as follows:

Enzyme	Disaccharide	Monosaccharides
Maltase	Maltose	Glucose + Glucose
Lactase	Lactose	Glucose + Galactose
Sucrase	Sucrose	Glucose + Fructose

>> **Intestinal lipase** targets fats, splitting them into fatty acids and glycerol.

The liver

The largest internal organ in the body, the liver (shown in Figure 14-5) is divided into a large right lobe and a small left lobe by the *falciform ligament*, another peritoneal fold. Two smaller lobes — the *quadrate* and *caudate* lobes — are found on the lower (inferior) and back (posterior) sides of the right lobe. The quadrate lobe surrounds and cushions the *gallbladder*, a pear-shaped structure that stores and concentrates *bile*, which it empties periodically through the *cystic duct* to the *common bile duct* and on into the duodenum during digestion. This is triggered by the hormone *cholecystokinin (CCK)*, which is released by cells in the duodenum. Bile aids in the digestion and absorption of fats; it consists of bile pigments, bile salts, and cholesterol.

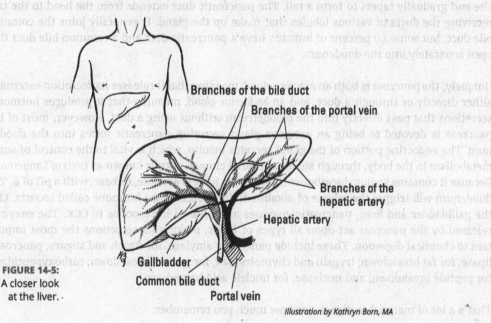

FIGURE 14-5:
A closer look
at the liver.

Illustration by Kathryn Born, MA

The liver secretes diluted bile through the right and left *hepatic ducts* into the common hepatic duct that joins the cystic duct coming from the gallbladder, forming the common bile duct. The cystic duct allows flow in both directions so bile is usually sent here. Liver tissue is made up of rows of cuboidal cells (see Chapter 5) separated by microscopic blood spaces called *sinusoids* creating the *hepatic lobules*, which are the functional units of the liver. Veins carrying the blood away from the intestines are merged through the mesentery into the *hepatic portal vein*. Thus, all absorbed nutrients are first sent to the liver to be metabolized. Blood from the hepatic artery releases oxygen into the sinusoids so that liver tissue can be oxygenated when processing the absorbed nutrients. The sinusoids also contain *Kupffer cells*, which phagocytize (eat) bacteria.

REMEMBER

Considering the number of vital roles the liver plays, its complexity isn't too surprising. Among the liver's various functions are

>> Production of blood plasma proteins, including *albumin,* antibodies to fend off disease, a blood anticoagulant called *heparin* that prevents clotting, and bile pigments from red blood cells: the yellow pigment *bilirubin* and the green bile pigment *biliverdin*

>> Storage of vitamins and minerals as well as glucose in the form of glycogen

» Conversion and utilization of fats, carbohydrates, and proteins

» Filtering and removal of nonfunctioning red blood cells, toxins, and waste products from amino acid breakdown, such as ammonia

The pancreas

Equally important, though not as large as the liver, the *pancreas* looks like a roughly 7-inch long, irregularly shaped prism. It has a broad head lodged in the curve of the duodenum. The head is attached to the body of the gland by a slight constriction called the neck, and the opposite end gradually tapers to form a tail. The pancreatic duct extends from the head to the tail, receiving the ducts of various lobules that make up the gland. It generally joins the common bile duct, but some 40 percent of humans have a pancreatic duct and a common bile duct that open separately into the duodenum.

REMEMBER

Uniquely, the pancreas is both an *exocrine gland,* meaning that it releases its secretion externally either directly or through a duct, and an *endocrine gland,* meaning that it produces hormonal secretions that pass directly into the bloodstream without using a duct. However, most of the pancreas is devoted to being an exocrine gland secreting pancreatic juices into the duodenum. The endocrine portion of the gland secretes insulin, which is vital to the control of sugar metabolism in the body, through small, scattered clumps of cells known as *islets of Langerhans.* Because it contains sodium bicarbonate, pancreatic juice is alkaline, or base, with a pH of 8. The duodenum will trigger the release of alkaline fluid by secreting a hormone called *secretin.* Like the gallbladder and liver, pancreatic enzymes are released in response to CCK. The enzymes released by the pancreas act upon all types of foods, making its secretions the most important to chemical digestion. These include pancreatic amylase, for starch and sugars; pancreatic lipase, for fat breakdown; trypsin and chymotrypsin, for protein breakdown; carboxypeptidase, for peptide breakdown; and nuclease, for nucleic acid breakdown.

That's a lot of material to digest. See how much you remember:

 Which of the following terms doesn't belong?

  a. Enterokinase

  b. Maltose

  c. Amylase

  d. Peptidase

  e. Sucrase

 Which of the following is NOT one of the liver's functions?

  a. Production of insulin

  b. Production of bile pigments

  c. Storage of vitamins and minerals

  d. Metabolism of absorbed nutrients

  e. Removal of old blood cells

47 Where are most nutrients absorbed?

    a. The pylorus

    b. The jejunum

    c. The ileum

    d. The fundus

    e. The duodenum

48 What role do villi play throughout the small intestine?

    a. They neutralize acids.

    b. They propel chyme along the intestinal route.

    c. They increase the surface area available to absorb nutrients.

    d. They perform mixing movements to increase access to nutrients.

    e. They further masticate the chyme.

49 What is the role of enterokinase in digestion?

    a. It activates a pancreatic enzyme that can break down protein.

    b. It acts as a proteolytic enzyme.

    c. It prompts the release of erepsin.

    d. It triggers enzyme release from the pancreas and liver.

    e. It inverts the functions of other enzymes.

50 What do Paneth and Kupffer cells have in common?

    a. They secrete mucus.

    b. They secrete digestive enzymes.

    c. They propel chyme through their corresponding organs.

    d. They both play a role in immunity.

    e. They share nothing in common.

51 When the small intestine receives the acidic chyme, the enterogastric reflex occurs causing:

a. the pancreas to release alkaline fluid.

b. the pyloric sphincter to close.

c. the ileocecal sphincter to open.

d. the jejunum to begin peristalsis.

e. the pancreas and gallbladder to release enzymes and bile.

One Last Look Before Leaving

Once chyme enters the large intestine through the *ileocecal sphincter*, it's our last chance to get anything out of it. What's left is packaged into *feces* and excreted from the body via *defecation*.

The appendix

Once considered a *vestigial organ* (one that has no known purpose), the *appendix* is found where the small and large intestine meet. It is true that this tiny, worm-like tube has no digestive function. However, it is comprised of lymphatic tissue and is thought to house our bacteria. This way, we can always repopulate the *intestinal flora* after a bout of diarrhea.

The large intestine

The large intestine, or colon, is about 5 ft. long proceeding up, across, and then down the peritoneal cavity. On its way out of the body, the chyme passes through the following:

Cecum → Ascending colon→ Transverse colon → Descending colon → Sigmoid colon → Rectum → Anus

The large intestine is about 3 inches wide at the start and decreases in width all the way to the rectum. The large intestine has no vili and secretes only mucus so no digestion occurs here. As the unabsorbed material moves through the large intestine, excess water is reabsorbed, drying out the material. In fact, most of the body's water absorption takes place in the large intestine bringing electrolytes along with it.

Bacteria that inhabit the large intestine feed off molecules that we cannot digest like cellulose (plant starch). They use this for energy, creating gas as a byproduct. Though the release of this gas, or *flatus*, can lead to some awkward social situations, the bacteria also synthesize vitamins (K, B12, thiamine, riboflavin) that we absorb through the mucosa. So it's a trade-off really. Perhaps when we pass gas, instead of saying "excuse me," we should say "thank you!"

Peristalsis moves the chyme through the large intestine slowly, segmenting it for maximum water absorption. Following a meal, a *mass movement* occurs where a strong wave of peristalsis pushes the newly formed feces toward the rectum. Once filled, stretch in the walls of the rectum initiates the *defecation reflex*. Thankfully, we can voluntarily inhibit this reflex by contracting the external anal sphincter. When we have prepared ourselves accordingly, we stop contracting the sphincter and allow the reflex to continue. This triggers one final peristaltic wave through the descending colon and feces is forced out of the rectum through the anus.

And that's the end of the line.

---

 52 True or False: The large intestine performs a role in digestion.

 53 Which of the following sphincters are under voluntary control?

    I. external anal sphincter

    II. internal anal sphincter

    III. upper esophageal sphincter

  a. I

  b. III

  c. I & III

  d. II & III

  e. I, II, & III

Answers to Questions on the Digestive Tract

The following are answers to the practice questions presented in this chapter.

1 Taking in food: **b. Ingestion.**

2 Elimination of waste: **e. Egestion.**

3 Movement of food from mouth to stomach: **c. Deglutition.**

4 Means of transporting food into the blood: **d. Absorption.**

5 Mechanical/chemical changing of food composition: **a. Digestion.**

6-16 Following is how Figure 14-1, the digestive system, should be labeled.

> 6. **c. Liver**; 7. **f. Gallbladder**; 8. **b. Large intestine**; 9. **g. Appendix**; 10. **e. Salivary glands**; 11. **i. Esophagus**; 12. **k. Stomach**; 13. **a. Pancreas**; 14. **d. Small intestine**; 15. **j. Rectum**; 16. **h. Anus**

17 Identify the correct sequence of the movement of food through the body: **a. Mouth → Pharynx → Esophagus → Stomach → Small intestine → Large intestine.**

TIP

Although remembering the sequence M-P-E-S-small-large can be helpful, you can also try this phrase to jog your memory: Most Phones Enable Speeches, from Small to Large.

18 Indicate the order of the layered walls of the alimentary canal, from the innermost proceeding outward. **Lumen; Mucosa; Submucosa; Muscularis, circular; Muscularis, longitudinal; Serosa**

WARNING

Be careful to note the order requested. A common pitfall is to give them backward and that is often an answer choice.

19-28 Following is how Figure 14-2, the tooth, should be labeled.

> 19. **c. Crown**; 20. **a. Neck**; 21. **g. Root**; 22. **f. Enamel**; 23. **b. Dentin**; 24. **i. Pulp cavity**; 25. **e. Gingiva**; 26. **j. Cementum**; 27. **d. Periodontal ligament**; 28. **h. Apical foramen**

29 What's the function of the soft palate, which vibrates when you snore? **d. It prevents food and liquid from getting into the nasal cavity.** So, as irritating as snoring may be, you need that soft palate to stay put.

30 What anchors the tongue? **c. The extrinsic muscles and the lingual frenulum.** Extrinsic means "outside," and anchor points occur outside, not inside (or intrinsically).

31 Which of the following statements about teeth is NOT true? **b. Each tooth has a single cuspid anchoring it.** You can rule out this answer option as false because a cuspid is a type of tooth, so it makes no sense that each tooth would have another type of tooth anchoring it.

32 If it hasn't been surgically removed, where can the palatine tonsil be found? **d. In the region where the soft palate meets the oropharynx.** Superior means above so choice e cannot be correct.

(33) Besides making it possible to spit, why do people have saliva? **e. All of the above.** Multifunctional stuff, that saliva. It facilitates swallowing, initiates the digestion of certain carbohydrates, moistens and lubricates the mouth and lips, as well as contains nifty lysosymes to destroy bacteria (the ones that have cell walls anyway).

(34-41) Following is how Figure 14-4, the stomach, should be labeled.

34. **b. Esophagus;** 35. **e. Serosa;** 36. **h. Longitudinal muscle layer;** 37. **a. Circular muscle layer;** 38. **f. Oblique muscle layer;** 39. **c. Rugae of the mucosa;** 40. **g. Pyloric sphincter;** 41. **d. Cardiac (lower esophageal) sphincter.**

(42) What does peristalsis do to food? **d. This sequential contraction of circular muscles helps move food through the esophagus.**

A bit of Greek may help you remember this term, which comes from the word peristaltikos, *which means "to wrap around."*

TIP

(43) Which part of the digestive tract isn't doing its job when you have heartburn? **d. The cardiac sphincter.** Biggest clue? Cardiac = heart.

(44) Where can chyme first be found along the digestive tract? **a. The pylorus.** Chyme is the thick, semiliquid mass of food that's ready to leave the stomach, so you can select the right answer if you remember your Greek prefixes and suffixes: *pyl–* means "gate," and *–orus* means "guard."

Here's a silly but effective memory tool for this term: When food is ready to leave the stomach, it rings a chime.

TIP

(45) Which of the following terms doesn't belong? **b. Maltose.** This is a sugar, whereas the other answer options are all enzymes (note the ending *–ase*).

(46) Which of the following is NOT one of the liver's functions? **a. Production of insulin.** That's what the pancreas does.

(47) Where are most nutrients absorbed? **b. The jejunum.** The duodenum does digestion (they both start with d) and the ileum is our last-ditch effort to get nutrients out, thus the jejunum does the bulk of the absorption.

To remember this one, keep in mind that each of the three sections of the small intestine plays a unique role. The duodenum neutralizes the stomach acid as the chyme enters the intestine (among other things), the jejunum absorbs most of the nutrients, and the ileum monitors for pathogens (or keeps you from getting "ill" — "il"eum. Get it?).

TIP

(48) What role do villi play throughout the small intestine? **c. They increase the surface area available to absorb nutrients.** They do other things along the way, but none of those functions are included in the other answers.

(49) What is the role of enterokinase in digestion? **a. It activates a pancreatic enzyme that can break down protein.**

It's tricky to remember which of these enzymes is inactive until it combines with something else. You can either try to memorize the function of each enzyme, or you can pick apart the terms. The prefix entero– *comes from the Greek word for "intestine." The suffix* –kinase *stems from the Greek word for "moving." "Moving through the intestine" sounds like a good guess, don't you think?*

TIP

50 What do Paneth and Kupffer cells have in common? **d. They both play a role in immunity.**

51 When the small intestine receives the acidic chyme, the enterogastric reflex occurs causing: **b. the pyloric sphincter to close.** While choice a seems reasonable, the release of alkaline fluid is triggered by the hormone secretin and not part of the reflex; the reflex merely controls how much chyme is allowed in at one time.

52 True or false: The large intestine performs a role in digestion. **False.** Every nutrient you're going to get out of the food you eat already has been absorbed by the time it reaches the large intestine. It's the body's primary location for reabsorbing water, but that's not a digestive role. Vitamins you absorb here were created by bacteria, not your digestive processes.

53 Which of the following sphincters are under voluntary control? **a. I.** Thankfully, the sphincter that controls the exit of feces (the external anal sphincter) is under our own control. The upper esophageal sphincter is made of skeletal muscle, leading you to believe that it could be voluntarily controlled, but its relaxation is part of the swallowing reflex and cannot be overridden.

Chapter **15**

Filtering Out the Junk: The Urinary System

I f you read Chapter 14 on the digestive system, you may be chewing on the idea that undigested food is the body's primary waste product. But it's not — that title belongs to urine. We make more of it than we do feces — in fact, our bodies are making small amounts of urine all the time — and we release it more often throughout the day. Most important, urine captures all the leftovers from our cells' metabolic activities and jettisons them before they can build up and become toxic. In addition, urine helps maintain the *homeostasis* of body fluids: electrolytes that control acid–base ratio, and the mixture of salt and water that regulates blood pressure.

In short, the urinary system

» Excretes useless and harmful material that it filters from blood plasma, including urea, uric acid, creatinine, and various salts

» Removes excess materials, particularly anything normally present in the blood that builds up to excessive levels

» Maintains proper osmotic pressure, or fluid balance, by eliminating excess water when concentration rises too high at the tissue level

» Reabsorbs water, glucose, ions, and amino acids

» Produces the hormones calcitriol (the hormonally active form of vitamin D that is necessary for calcium absorption) and erythropoietin (which stimulates red blood cell production in the bone marrow), as well as the hormone renin (that works to maintain blood pressure)

This chapter explains how the urinary system collects, manages, and excretes the waste that the body's cells produce as they go about busily metabolizing all day. You discover the parts of the kidneys, ureter, urinary bladder, and urethra.

Examining the Kidneys, the Body's Filters

The kidneys are nonstop filters that sift through 1.2 liters of blood per minute. Humans have a pair of kidneys just above the waist (lumbar region) toward the back of the abdominal cavity. Although sometimes the same size, the left kidney tends to be a bit larger and more superior than the right due to the size of the liver. The last two pairs of ribs surround and protect each kidney, and a layer of fat, called *perirenal (perinephric) fat*, provides additional cushioning. Kidneys are *retroperitoneal*, which means that they're posterior to the parietal peritoneum. The *renal capsule*, or outer lining of the kidney, is a layer of connective tissue full of collagen fibers; these fibers extend outward to anchor the organ to surrounding structures.

Kidney structure

Each kidney is dark red, about 4½ inches long, and shaped like a bean (hence the type of legumes called kidney beans). The portion of the bean that folds in on itself, referred to as the *medial border*, is concave with a deep depression in it called the *hilum*. The hilum opens into a fat-filled space called the *renal sinus*, which in turn contains the *renal pelvis*, *renal calyces*, blood vessels, and nerves. The *renal artery* and *renal vein*, which provide the kidney's blood supply, as well as the ureter that carries urine to the bladder, leave the kidney through the hilum.

Immediately below the renal capsule is a granular layer called the *renal cortex*, and just below that is an inner layer called the *medulla* that folds into anywhere from 8 to 18 conical projections called the *renal pyramids*. Between the pyramids are *renal columns* that extend from the cortex inward to the renal sinus. The columns allow for passage of the blood vessels. The tips of the pyramids, the *renal papillae*, empty their contents into a collecting area called the *minor calyx*. It's one of several saclike structures referred to as the *minor* and *major calyces*, which form the start of the urinary tract's "plumbing" system and collect urine transmitted through the papillae of the pyramids. Although the number varies between individuals, a single minor calyx surrounding the papilla of one pyramid combines into four or five minor calyces, which merge into two or three major calyces. Urine passes through the minor calyx into its major calyx, through the pelvis, and then into the ureter for the trip to the bladder.

Going microscopic

At the microscopic level, each kidney contains more than one million tiny tubes known as *nephrons*. These are the primary functional units of the urinary system. At one end, each nephron is closed off and folded into a small double-cupped structure called a *Bowman's capsule*, or the *glomerular capsule*, where the actual process of filtration occurs. Leading away from the capsule, the nephron forms into the *proximal convoluted tubule* (PCT), which is lined with cuboidal epithelial cells having microvilli brush borders that increase the area of absorption. This tube straightens to form a structure called the *descending loop of Henle* and then bends back in a hairpin turn into another structure called the *ascending loop of Henle*. After that, the tube becomes convoluted again, forming the *distal convoluted tubule* (DCT), which is made of the same types of cells as the first, or proximal, convoluted tubule but without any microvilli. This tubule connects to a *collecting duct* that it shares with the output ends of many other nephrons. The collecting ducts lie within the pyramids.

Because of their role as the body's key filters, the kidneys receive about 20 percent of all the blood pumped by the heart each minute. A large branch of the abdominal aorta, called the *renal artery*, carries that blood to the kidneys. After branching into smaller and smaller vessels, the blood eventually moves through interlobular arteries to the *afferent arterioles*, each of which branches into tufts of five to eight capillaries called a *glomerulus* (the plural is *glomeruli*) inside the *glomerular (Bowman's) capsule*. In the glomerulus, pressure differences force filtration of solutes, fluids, and other glomerular filtrates through the capillary walls into the glomerular (Bowman's) capsule. The glomerular capillaries come back together to form *efferent arterioles*, which then branch to form the *peritubular capillaries* surrounding the convoluted tubules, the loop of Henle, and the collecting duct. The capillaries come together once again to form a small vein (*venule*) that empties blood into the interlobular vein, eventually moving the blood into the renal vein to depart the kidneys.

Each glomerulus and its surrounding glomerular capsule make up a single *renal corpuscle* where filtration takes place. Like all capillaries, glomeruli have thin, membranous walls, but unlike their capillary cousins elsewhere, these vessels have unusually large pores called *fenestrations* or *fenestrae* (from the Latin word *fenestra* for "window"). Plasma is pushed out here but gas exchange does not take place, which is why the glomerular capillaries merge into an efferent arteriole (as opposed to a vein). Normal capillary exchange takes place in the peritubular capillaries.

Are you getting the flow of all of this? Let's try some questions.

 Use the following terms to identify the parts of the kidney in Figure 15-1.

a. Renal vein

b. Cortex

c. Ureter

d. Renal pelvis

e. Renal artery

f. Medulla (renal pyramids)

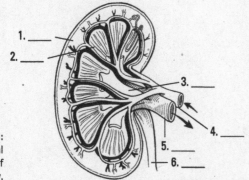

FIGURE 15-1:
Internal
anatomy of
the kidney.

Illustration by Kathryn Born, MA

7–14 Use the following terms to identify the nephron structures in Figure 15-2.

a. Loop of Henle

b. Glomerular (Bowman's) capsule

c. Collecting duct

d. Proximal convoluted tubule

e. Interlobular vein

f. Peritubular capillary bed

g. Distal convoluted tubule

h. Interlobular artery

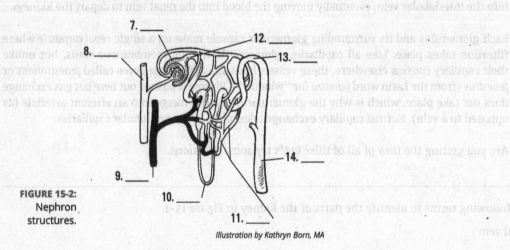

7. ____

8. ____

12. ____

13. ____

9. ____

10. ____

11. ____

14. ____

FIGURE 15-2:
Nephron
structures.

Illustration by Kathryn Born, MA

15–19 Match the anatomical terms with their descriptions.

a. Composed of folds forming the renal pyramids and the columns in between

b. Granular outer layer

c. Irregular, saclike structures for collecting urine in the renal pelvis

d. Transports urine from the nephrons

e. Funnels urine into the ureter

15 ___B___ Cortex

16 ___A___ Medulla

17 ___E___ Renal pelvis

18 ___C___ Calyx

19 ___D___ Collecting duct

20 The human kidney lies outside the abdominal cavity. This makes it

    a. retroperitoneal.

    b. parietal.

    c. visceral.

    d. endocoelomic.

    e. exterocoelomic.

21 What is the primary functional unit of a kidney?

    a. The medulla

    b. The loop of Henle

    c. The renal papillae

    d. The nephron

    e. The renal sinus

22 Where are brush borders of villi found primarily?

    a. Ascending loop of Henle

    b. Proximal convoluted tubule

    c. Glomerular (Bowman's) capsule

    d. Descending loop of Henle

23 Although the kidney has many functional parts, where does filtration primarily occur?

    a. Inside the distal convoluted tubules

    b. Outside the loop of Henle

    c. Just beyond the renal cortex

    d. In conjunction with the renal sinus

    e. Within the glomerulus inside a glomerular (Bowman's) capsule

Focusing on Filtering

TIP

To understand how the renal corpuscles work, think of an espresso machine: Water is forced under pressure through a sieve containing ground coffee beans, and a filtrate called brewed coffee trickles out the other end. Something similar takes place in the renal corpuscles.

Hydrostatic pressure forces plasma through the holes in the glomerulus, which is referred to as *glomerular filtration*. This filtrate, to the tune of about 125 mL/min., is collected by the surrounding glomerular capsule and ushered into the PCT (see Figure 15-3). So we have a rather effective filter but we can't afford to lose that much water; we couldn't possibly drink fast enough to replace it plus we'd urinate constantly. Thus, we have to get most of that water back into the plasma (blood flow). And we do, at 124 mL/min. on average. Furthermore, there are molecules in the filtrate that we'd rather not get rid of such as ions and glucose. That's the job of the nephron — to get back everything in the filtrate that is useful to us.

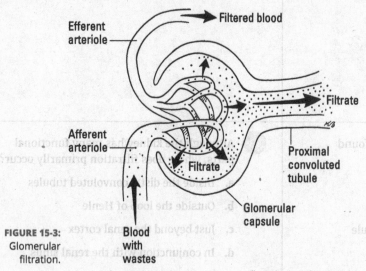

FIGURE 15-3: Glomerular filtration.

Illustration by Kathryn Born, MA

Retaining water

The work of the nephron takes place in three phases, each corresponding to a different section which you can see illustrated in Figure 15-4.

TIP

Keep in mind your perspective when reading through the actions of the nephron. In the end, we don't care about what's in the urine, we care about what's in (or isn't in) the blood. So when we say reabsorption, we're talking about reabsorbing back into the blood stream. When we say secretion, we're talking about secreting from the bloodstream.

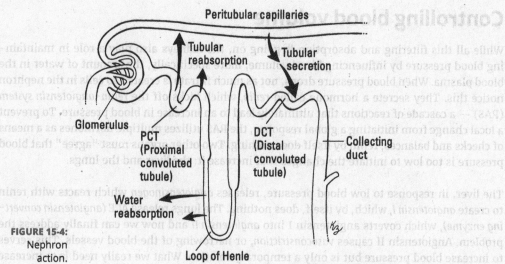

FIGURE 15-4:
Nephron
action.

Peritubular capillaries

Tubular
reabsorption

Tubular
secretion

Glomerulus

PCT
(Proximal
convoluted
tubule)

DCT
(Distal
convoluted
tubule)

Collecting
duct

Water
reabsorption

Loop of Henle

Illustration by Kathryn Born, MA

» **Tubular reabsorption** primarily occurs in the PCT. Molecules that were dissolved in the plasma are now in the filtrate and this is our chance to get them back. Glucose is 100-percent reabsorbed here along with various proteins, amino acids, ions, and of course, water. These molecules are transported out of the filtrate and into the interstitial fluid surrounding the nephron, where they can diffuse into the peritubular capillaries, thus back into blood flow. Wastes such as *urea* and *uric acid* remain in the filtrate.

» **Water reabsorption** primarily occurs in the loop of Henle. Unfortunately, we don't have a mechanism to directly transport water. We can, however, pump ions which can lead to the movement of water by osmosis. Along the ascending loop there are sodium pumps which transport Na^+ out of the filtrate. Anions such as Cl^- follow the sodium ions out, attracted to the opposing charge. In the interstitial fluid, these ions bond forming salts. Consequently, the solute concentration increases — which means the water concentration decreases. Since the concentration of water is now lower outside of the nephron, it moves out. Since so much was pushed out during glomerular filtration, the concentration of water is even lower inside the peritubular capillaries so water flows back into the bloodstream. This is called *countercurrent exchange* and ensures that we are always getting water out of the filtrate by controlling the solute concentration of the interstitial fluid.

» **Tubular secretion** primarily occurs in the DCT. There are two types of wastes that remain in the blood of the peritubular capillaries — those that were too numerous to all be pushed out during filtration (such as hydrogen ions) and those that were too big (such as histamine and drug compounds). Both are targeted by active transport pumps pulling them out of bloodflow and into the filtrate just before it enters the collecting duct.

Though each nephron action primarily occurs to a specific section, it's important to realize that many of them — especially water reabsorption — can occur anywhere.

WARNING

Controlling blood volume

While all this filtering and absorption is going on, the kidneys also play a role in maintaining blood pressure by influencing blood volume; more specifically, the amount of water in the blood plasma. When blood pressure drops, not as much filtrate is created so cells in the nephron notice this. They secrete a hormone called *renin*, which kicks off the *renin-angiotensin system* (RAS) — a cascade of reactions that ultimately lead to an increase in blood pressure. To prevent a local change from initiating a global response, the RAS utilizes multiple hormones as a means of checks and balances; renin by itself does nothing. Two other organs must "agree" that blood pressure is too low to initiate the changes that increase it: the liver and the lungs.

The liver, in response to low blood pressure, releases *angiotensinogen* which reacts with renin to create *angiotensin I*, which, by itself, does nothing. The lungs release *ACE (angiotensin converting enzyme)*, which coverts angiotensin I into *angiotensin II* and now we can finally address the problem. Angiotensin II causes *vasoconstriction*, or narrowing of the blood vessels. This serves to increase blood pressure but is only a temporary solution. What we really need is to increase the blood volume, which we accomplish by retaining more water in the kidneys. In addition to feeling thirsty (thus prompting you to help the cause by drinking more water), angiotensin II triggers the posterior pituitary to release *ADH (anti-diuretic hormone)* and stimulate the adrenal cortex to release *aldosterone*, both of which head to the kidneys. Aldosterone triggers sodium reabsorption in the DCT, while ADH does so in the collecting duct (which is our last chance to get anything out of the filtrate). Water follows sodium and voilà! More water in the bloodstream, low blood pressure problem solved.

You've absorbed a lot in the last few paragraphs. See how much of it is getting caught in your filters.

---

 24 The correct sequence for removal of material from the blood through the nephron is

a. afferent arteriole → glomerulus → proximal convoluted tubule → loop of Henle → distal convoluted tubule → collecting duct.

b. afferent arteriole → glomerulus → distal convoluted tubule → loop of Henle → proximal convoluted tubule → collecting duct.

c. afferent arteriole → collecting duct → glomerulus → proximal convoluted tubule → loop of Henle → distal convoluted tubule.

d. efferent arteriole → proximal convoluted tubule → glomerulus → loop of Henle → distal convoluted tubule → collecting duct.

e. efferent arteriole → glomerulus → proximal convoluted tubule → loop of Henle → distal convoluted tubule → collecting duct.

25 If 125 milliliters of material is captured per minute in the glomerular (Bowman's) capsules, why is only 1 milliliter of urine generated each minute?

    a. The renal pelvis siphons off most of it.

    b. A series of tubules allow most of it to be reabsorbed into the bloodstream.

    c. The material is highly concentrated by the time it becomes urine.

    d. Hydrostatic pressure forces most of the material back out again.

26-29 Match the event to the location in the nephron where it is most likely to occur.

    a. renal corpuscle

    b. PCT

    c. DCT

    d. loop of Henle

    26 _____ Sodium is actively transported

    27 _____ Toxins are secreted

    28 _____ Plasma is pushed out

    29 _____ Glucose is reabsorbed

30 True or False: Countercurrent exchange occurs when water is actively transported out of the nephron and into the interstitial fluid where it can then passively enter the peritubular capillaries.

31-35 Match the hormone to its origin.

    a. posterior pituitary

    b. lungs

    c. liver

    d. kidney

    e. adrenal cortex

    31 _____ Renin

    32 _____ Angiotensinogen

    33 _____ Aldosterone

    34 _____ ADH

    35 _____ ACE

36 Which hormone is responsible for water retention in the kidney from the last place possible before the filtrate becomes urine?

    a. renin

    b. angiotensin

    c. ACE

    d. aldosterone

    e. ADH

Getting Rid of the Waste

After your kidneys filter out the junk, it's time to deliver it to the bladder and the urethra via the ureters. The following sections explain how that's done.

Surfing the ureters

Ureters are narrow, muscular tubes through which the collected waste travels. About 10 inches long, each ureter descends from a kidney to the posterior lower third of the bladder. Like the kidneys themselves, the ureters are behind the peritoneum outside the abdominal cavity, so the term *retroperitoneal* applies to them, too. The inner wall of the ureter is a simple mucous membrane composed of elastic transitional epithelium. It also has a middle layer of smooth muscle tissue that propels the urine by peristalsis — the same process that moves food through the digestive system. So rather than trickling into the bladder, urine arrives in small spurts as the muscular contractions force it down. The tube is surrounded by an outer layer of fibrous connective tissue that supports it during peristalsis.

Ballooning the bladder

The urinary *bladder* is a large, muscular bag that lies in the pelvis behind the pubis bones. In human females, it's beneath the uterus and in front of the vagina (see Chapter 17). In human males, the bladder lies between the rectum and the symphysis pubis (see Chapter 16). There are three openings in the bladder: two on the back side where the ureters enter and one on the front for the *urethra*, the tube that carries urine outside the body (see the next section). The bladder's *trigone* is the triangular area between these three openings. The *neck* of the bladder surrounds the urethral attachment, and the *internal sphincter* (smooth muscle that provides involuntary control) encircles the junction between the urethra and the bladder. The external urethral sphincter surrounds the urethra nearer to its orifice and this one, thankfully, we have voluntary control over.

Inside, the bladder is lined with highly elastic transitional epithelium (described in Chapter 5). When full, the bladder's lining is smooth and stretched; when empty, the lining lies in a series of folds called *rugae* (just as the stomach does). When the bladder fills, the increased pressure stimulates the organ's stretch receptors, particularly those in the trigone, and initiates the *micturition reflex* (which you feel as the need to urinate). The bladder can hold 600–800 milliliters of urine, but usually the pressure receptor's response will cause it to empty before reaching maximum capacity. (See the later section "Spelling relief: Urination" for more details.)

Distinguishing the male and female urethras

Both males and females have a *urethra*, the tube that carries urine from the bladder to a body opening, or orifice. Both males and females have an internal sphincter controlled by the autonomic nervous system and composed of smooth muscle tissue to guard the exit from the bladder. Both males and females also have an external sphincter composed of skeletal muscle tissue that's under voluntary control. But as we all well know, the exterior plumbing is rather different.

The female urethra is about one and a half inches long and lies close to the vagina's anterior (front) wall. It opens just anterior to (in front of) the vaginal opening. The external sphincter for the female urethra lies just inside the urethra's exit point.

The male urethra is about 8 inches long and carries a different name as it passes through each of three regions:

>> The *prostatic urethra* leaving the bladder contains the internal sphincter and passes through the prostate gland. Several openings appear in this region of the urethra, including a small opening where sperm from the vas deferens and ejaculatory duct enters, and prostatic ducts where fluid from the prostate enters.

>> The *membranous urethra* is a small 1- or 2-centimeter portion that contains the external sphincter and penetrates the pelvic floor. The tiny *bulbourethral* (or *Cowper's*) *glands* lie on either side of this region.

>> The *cavernous urethra*, also known as the *spongy*, or *penile urethra*, runs the length of the penis on its ventral surface through the *corpus spongiosum*, ending at a vertical slit at the end of the penis.

Spelling relief: Urination

Urination, known by the proper term *micturition*, occurs when the bladder is emptied through the urethra. Although urine is created continuously, it's stored in the bladder until the individual finds a convenient time to release it. Mucus produced in the bladder's lining protects its walls from any acidic or alkaline effects of the stored urine. When there is about 200 milliliters of urine distending the bladder walls, stretch receptors transmit impulses to warn that the bladder is filling. *Afferent* impulses are transmitted to the spinal cord, and *efferent* impulses return to the bladder, forming a reflex arc that causes the internal sphincter to relax and the *detrusor muscle* of the bladder to contract, forcing urine into the urethra. The afferent impulses continue up the spinal cord to the brain, creating the urge to urinate. Because the external sphincter is composed of skeletal muscle tissue, no urine usually is released until the individual voluntarily opens the sphincter.

Now test your knowledge of how the human body gets rid of its waste:

EXAMPLE

Q. The separation of the reproductive and urinary systems is complete in the human

a. male.

b. female.

c. male and female.

A. The correct answer is female. The male urethra runs through the same "plumbing" as the male reproductive system.

37 How can the interior of the bladder expand as much as it does?

a. Its ciliated columnar epithelium can shift away from each other.

b. It is lined with transitional epithelium.

c. Its cuboidal epithelium can realign its structure as needed.

d. Its white fibrous connective tissue gives way under pressure.

 38 How does urine move from the kidney down the ureter?

a. By fibrillation

b. By flexure

c. Simply because of gravity

d. By peristaltic contractions

e. It simply flows through.

 39 The external sphincter of the urinary tract is made up of

a. circular striated (skeletal) muscle.

b. rugae.

c. simple mucous membrane.

d. the membranous urethra.

e. smooth muscle.

40 What is the function of the internal sphincter at the junction of the bladder neck and the urethra?

a. To stimulate the expulsion of urine from the bladder

b. To keep foreign substances and infections from entering the bladder

c. To prevent urine leakage from the bladder

d. To prevent return of urine from the urethra back into the bladder

e. To relax when voluntarily triggered to expel urine

41 True or False: Urine production occurs sporadically throughout the day.

Answers to Questions on the Urinary System

The following are answers to the practice questions presented in this chapter.

1-6 Following is how Figure 15-1, the internal anatomy of the kidney, should be labeled.

1. **b. Cortex**; 2. **f. Medulla (renal pyramids)**; 3. **d. Renal pelvis**; 4. **e. Renal artery**; 5. **a. Renal vein**; 6. **c. Ureter**

7-14 Following is how Figure 15-2, the nephron, should be labeled.

7. **b. Glomerular (Bowman's) capsule**; 8. **h. Interlobular artery**; 9. **e. Interlobular vein**; 10. **a. Loop of Henle**; 11. **f. Peritubular capillary bed**; 12. **d. Proximal convoluted tubule**; 13. **g. Distal convoluted tubule**; 14. **c. Collecting duct**

15 Cortex: **b. Granular outer layer**

16 Medulla: **a. Composed of folds forming the renal pyramids and the columns in between**

17 Renal pelvis: **e. Funnels urine into the ureter**

18 Calyx: **c. Irregular, saclike structures for collecting urine in the renal pelvis**

19 Collecting tubule: **d. Transports urine from the nephrons**

20 The human kidney lies outside the abdominal cavity. This makes it **a. retroperitoneal.** *Peritoneal* refers to the peritoneum, the membrane lining the abdominal cavity; and *retro* can be defined as "situated behind."

21 What is the primary functional unit of a kidney? **d. The nephrons.** Each nephron contains a series of the parts needed to do the kidney's filtering job.

22 Where are brush borders of villi found primarily? **b. Proximal convoluted tubule.** Those brush borders provide extra surface area for reabsorption, so it makes sense that they congregate in the first area after filtration.

23 Although the kidney has many functional parts, where does filtration primarily occur? **e. Within the glomerulus inside a glomerular (Bowman's) capsule.** The glomerulus is a collection of capillaries with big pores, so think of it as the initial filtering sieve.

24 The correct sequence for removal of material from the blood through the nephron is **a. afferent arteriole → glomerulus → proximal convoluted tubule → loop of Henle → distal convoluted tubule → collecting tubule.** In short, blood comes through the artery (arteriole) and material gloms onto the nephron before twisting through the near (proximal) tubes, looping the loop, twisting through the distant (distal) tubes, and collecting itself at the other end. Try remembering artery-glom-proxy-loop-distant-collect.

25 If 125 milliliters of material is captured per minute in the glomerular (Bowman's) capsules, why is only 1 milliliter of urine generated each minute? **b. A series of tubules allow most of it to be reabsorbed into the bloodstream.** Remember the subtraction puzzle we discuss earlier in this chapter? We count up 124 milliliters of reabsorption.

26 Sodium is actively transported: **d. loop of Henle**

(27) Toxins are secreted: **c. DCT**

(28) Plasma is pushed out: **a. renal corpuscle.** Corpuscle is the term for the glomerulus and the Bowman's capsule together.

(29) Glucose is reabsorbed: **b. PCT**

(30) Countercurrent exchange occurs when water is actively transported out of the nephron and into the interstitial fluid, where it can then passively enter the peritubular capillaries. **False.** We can't move water, we can only encourage it to move. To do this, we actively pump Na^+ out of the nephrons, which leads to the formation of salts in the interstitial fluid. Increased solute concentration means decreased water concentration so water exits the nephron passively.

(31) Renin: **d. kidney**

(32) Angiotensinogen: **c. liver**

(33) Aldosterone: **e. adrenal cortex**

(34) ADH: **a. posterior pituitary**

(35) ACE: **b. lungs**

(36) Which hormone is responsible for water retention in the kidney from the last place possible before the filtrate becomes urine? **e. ADH.** The last place possible is the collecting duct which is where ADH triggers ion pumps.

(37) How can the interior of the bladder expand as much as it does? **b. It is lined with transitional epithelium.** It's transitional because it needs to be able to stretch and collapse as needed.

(38) How does the urine move from the kidney down the ureter? **e. By peristaltic contractions.** It's the same action that moves food through the digestive system.

(39) The external sphincter of the urinary tract is made up of **a. circular striated (skeletal) muscle.** The external sphincter is under a person's control. The only tissue in this list of choices that features voluntary control is skeletal muscle.

(40) What is the function of the internal sphincter at the junction of the bladder neck and the urethra? **c. To prevent urine leakage from the bladder.** The sphincter's makeup of smooth muscle tissue helps it do so, which means it's not under voluntary control.

(41) Urine production occurs sporadically throughout the day. **False.** The kidneys produce urine continuously in a healthy individual.

5

Survival of the Species

Chapter 16

Why Ask Y? The Male Reproductive System

ndividually, humans don't need to reproduce to survive. But to survive as a species, a number of individuals must produce and nurture a next generation, carrying their uniqueness forward in the gene pool. Humans are born with the necessary organs to do just that.

In this chapter, you get an overview of the parts and functions of the male reproductive system, along with plenty of practice questions to test your knowledge. (We cover the guys first because their role in the basic reproduction equation isn't nearly as long or complex as that of their mates. We address the female reproductive system in Chapter 17.)

Identifying the Parts of the Male Reproductive System

On the outside, the male reproductive parts, which you can see in Figure 16-1, are straightforward — a *penis* and a *scrotum*. At birth, the apex of the penis is enclosed in a fold of skin called the *prepuce*, or *foreskin*, which often is removed during a procedure called *circumcision*.

The scrotum is a pouch of skin divided in half on the surface by a ridge called a *raphe* that continues up along the underside of the penis and down all the way to the anus. The left side of the scrotum tends to hang lower than the right side to accommodate a longer *spermatic cord*, which we explain later in this section.

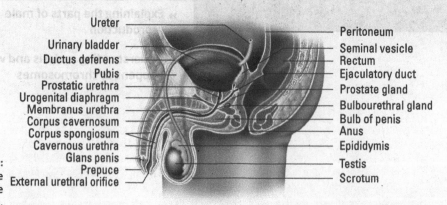

Ureter
Urinary bladder
Ductus deferens
Pubis
Prostatic urethra
Urogenital diaphragm
Membranus urethra
Corpus cavernosum
Corpus spongiosum
Cavernous urethra
Glans penis
Prepuce
External urethral orifice

Peritoneum
Seminal vesicle
Rectum
Ejaculatory duct
Prostate gland
Bulbourethral gland
Bulb of penis
Anus
Epididymis
Testis
Scrotum

FIGURE 16-1:
The male
reproductive
system.

Illustration by Imagineering Media Services Inc.

There are two scrotal layers: the *integument*, or outer skin layer, and the *dartos tunic*, an inner smooth muscle layer that contracts when cold and elongates when warm. Why? That has to do with the two *testes* (the singular is *testis*) inside (see Figure 16-2). These small ovoid glands, also referred to as *testicles*, need to be a bit cooler than body temperature in order to produce viable sperm for reproduction. When the dartos tunic becomes cold, such as when a man is swimming, it contracts and draws the testes toward the body for warmth. When the dartos tunic becomes overly warm, it elongates to allow the testes to hang farther away from the heat of the body.

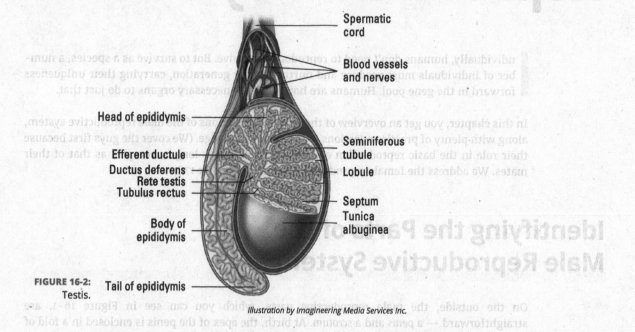

Spermatic
cord

Blood vessels
and nerves

Head of epididymis

Efferent ductule
Ductus deferens
Rete testis
Tubulus rectus

Body of
epididymis

Seminiferous
tubule

Lobule

Septum
Tunica
albuginea

FIGURE 16-2:
Testis.

Tail of epididymis

Illustration by Imagineering Media Services Inc.

A fibrous capsule called the *tunica albuginea* encases each testis and extends into the gland forming incomplete *septa* (partitions), which divide the testis into about 200 *lobules*. These compartments contain small, coiled *seminiferous tubules*. *Spermatogenesis*, or production of sperm cells, takes place in the seminiferous tubules via meiosis. This occurs in the walls of the seminiferous tubules with the newly created gametes, called *spermatids*, beginning their development. By the time they are released into the lumen, now called *spermatozoa*, they are not fully developed, still having a rounded head and a short tail. Spermatogenic (sperm-forming) cells are produced by a series of cellular divisions. These cells, called *spermatogonia*, divide continually by mitosis until puberty. (You can find out more about mitosis in Chapter 4.) Spermatogenesis begins during puberty when a spermatogonium divides mitotically to produce two types of daughter cells:

>> **Type A cells** remain in the tubule, producing more spermatogonia.

>> **Type B cells** undergo meiosis to produce four sperm, which we review in the next section.

Cells of the seminiferous tubules, called *Sertoli cells* (also known as *sustentacular cells* or *nurse cells*), are activated by follicle stimulating hormone (FSH), which is synthesized and secreted by the anterior pituitary gland. (See Chapter 10 for more information on these hormones.) The sustentacular cells secrete androgen-binding protein that concentrates the male sex hormone *testosterone* close to the germ cells to maintain the environment the spermatogonia need to develop and mature.

Distributed in gaps between the tubules are *Leydig cells* that produce testosterone. The tubules of each lobule come together in an area called the *mediastinum testis* and straighten into *tubuli recti* before forming a network called the *rete testis* that leads to the *efferent ducts* (also called *ductules*). These ducts carry sperm to an extremely long (about 20 feet), tightly coiled tube called the *epididymis* for storage and maturation of the sperm. This is where sperm fully mature, developing their flagella (though they are not yet able to swim; that does not happen until after ejaculation) and an *acrosome*, the outer casing around the head that contains enzymes to aid in fertilization.

The epididymis merges with the *ductus deferens*, or *vas deferens*, which carries sperm up into the spermatic cord, which also encases the testicular artery and vein, lymphatic vessels, and nerves. Convoluted pouches called *seminal vesicles* lie behind the base of the bladder and secrete an alkaline fluid containing fructose, vitamins, and amino acids to nourish sperm as they enter the *ejaculatory duct* (refer to Figure 16-1). The seminal vesicle also contributes prostaglandins (a hormone) to the semen to trigger muscle contraction once in the female reproductive tract, aiding the sperm's journey toward the ultimate prize.

From there the mixture containing sperm enters the *prostatic urethra* that's surrounded by the *prostate gland.* This gland secretes a thin, opalescent substance that precedes the sperm in an ejaculation, contributing approximately 30 percent of the semen. The alkaline nature of this substance effectively neutralizes the acidity found in the male urethra and reduces the natural acidity of the female's vagina to prepare it to receive the sperm. Two yellowish pea-sized bodies called *bulbourethral glands,* or *Cowper's glands,* lie on either side of the urethra and secrete a clear alkaline lubricant prior to ejaculation; it neutralizes the acidity of the urethra and acts as a lubricant for the penis. Once all the glands have added protective and nourishing fluids to the 400 to 500 million departing sperm, the mixture is known as *seminal fluid* or *semen.*

During sexual arousal, physical and/or mental stimulation causes the brain to send chemical messages via the central nervous system to nerves in the penis telling its blood vessels to dilate so that blood can flow freely into the penis. A *parasympathetic reflex* is triggered that stimulates arterioles to dilate in three columns of spongy erectile tissue in the penis — the *corpus spongiosum* surrounding the urethra and the two *corpora cavernosa,* on the dorsal surface of the erect penis. As the arterioles dilate, blood flow increases while vascular shunts constrict and reroute the blood supply from the bottom to the topside of the penis. Swelling with blood, the penis further compresses its vascular drainage, resulting in a rigid erection that makes the penis capable of entering the female's vagina. Most of the rigidity of the erect penis results from blood pressure in the corpora cavernosa; high blood pressure in the corpus spongiosum could close the urethra, preventing passage of the semen.

Once a sufficient level of stimulation has been received, the *sympathetic nervous system* triggers the two-part process of release. First, rhythmic contractions in the epididymis force sperm through the seminal vesicle and into the prostate gland where they both add their secretions. Then, muscles in the prostate propel the semen into the urethra. This is the process of *emission*. Next, through continuing sympathetic impulse, muscles along the shaft of the penis contract to propel the semen out of the urethra. Once the fluid exits out of the *urethral orifice* (the slit on the *glans penis*, or head of the penis), *ejaculation* has occurred.

See how familiar you are with the male anatomy by tackling these practice questions:

---

1 Why does the left side of the scrotum tend to hang lower than the right?

   a. To adapt to changes occurring during puberty

   b. Because the left side contains the larger testicle

   c. To accommodate a longer spermatic cord

   d. Because most men are right-handed

   e. It doesn't, they hang evenly

2 The _____ is/are the source of the action when the scrotum adjusts to surrounding temperatures.

   a. testes

   b. bulbourethral (Cowper's) glands

   c. dartos tunic

   d. spermatic cord

   e. prostatic urethra

3-8 Fill in the blanks in the following sentences.

Sperm-forming cells, or 3._____, divide throughout the reproductive lifetime of the male by a process called 4._____. But after puberty, these cells begin to produce two types of 5._____ cells. Type A cells remain in the wall of the 6._____, producing more of themselves. Type B cells produce 7._____ through the process of 8._____

9 Where is testosterone produced?

   a. Seminiferous tubules

   b. Interstitial (Leydig) cells

   c. Prostate gland

   d. Sertoli (Subtentacular or nurse) cells

   e. Epididymis

10 Select the correct sequence for the movement of sperm:

    a. Tubuli recti → rete testis → seminiferous tubules → epididymis → ductus deferens → urethra

    b. Seminiferous tubules → tubuli recti → rete testis → ejaculatory duct → epididymis → ductus deferens → urethra

    c. Epididymis → ejaculatory duct → tubuli recti → rete testis → seminiferous tubules → ductus deferens → urethra

    d. Seminiferous tubules → ejaculatory duct → tubuli recti → rete testis → epididymis → ductus deferens → urethra

    e. Seminiferous tubules → tubuli recti → rete testis → epididymis → ductus deferens → ejaculatory duct → urethra

11 As the sperm moves through the reproductive tract, which of the following does not add a secretion to it?

    a. Leydig (interstitial) cells

    b. Cowper's (bulbourethral) glands

    c. Seminal vesicle

    d. Prostate

12 Sperm storage and development takes place in the convoluted tube called the

    a. seminiferous tubule.

    b. rete testis.

    c. spermatic cord.

    d. scrotum.

    e. epididymis.

13 After all the proper glands have secreted fluids to nourish and protect the departing sperm, the substance ejaculated is called

    a. stroma.

    b. semen.

    c. prepuce.

    d. spermatozoa.

    e. inguinal.

14 An average ejaculation will contain about
    _____ sperm.

   a. 40 to 50 million

   b. 400 to 500 million

   c. 400 to 500

   d. 400,000 to 500,000

   e. 4 to 5 million

15 During sexual arousal, the penis becomes
    hard and erect. What is happening to
    cause that?

   a. Muscles in the base of the penis are con-
       tracting while other muscles toward the
       end of the penis are expanding.

   b. The corpus spongiosum and the corpora
       cavernosa draw slowly away from one
       another.

   c. Blood vessels are dilating to allow free
       flow of blood into the penis, while vascu-
       lar shunts constrict drainage and cause
       blood pressure to rise in the corpora
       cavernosa.

   d. The Cowper's glands are squeezing addi-
       tional fluids into the shaft of the penis.

   e. A sympathetic impulse triggers the blood
       vessels to close, trapping blood in the
       corpus spongiosum.

16 Explain the difference between emission and ejaculation.

_____

_____

_____

_____

Packaging the Chromosomes for Delivery

Sperm, the male sex cells, are produced during a process called *meiosis* (which also produces the female sex cell, or *ovum*). Meiosis involves two divisions:

>> **Reduction division (meiosis I):** The first, a *reduction division,* divides a single *diploid* cell with two sets of chromosomes into two *haploid* cells with only one set each. These haploid cells have two sets of alleles for each chromosome, though, so another division must occur.

>> **Division by mitosis (meiosis II):** The second process is a division by *mitosis* that divides the two haploid cells into four cells with a single set of chromosomes, each with one allele.

Review the reproduction terms in Table 16-1.

Table 16-1 Reproduction Terms to Know

Terms That Vary by Gender	General Term	Male Term	Female Term
Sex organs	Gonads	Testes	Ovaries
Original cell	Gametocyte	Spermatocyte	Oocyte
Meiosis	Gametogenesis	Spermatogenesis	Oogenesis
Sex cell	Gamete	Sperm or Spermatozoa	Ovum

REMEMBER

A *diploid cell* (or 2N) has two sets of chromosomes, whereas a *haploid cell* (or 1N) has one set of chromosomes. An allele is the variety, like instructions for attached earlobes versus instructions for free ones.

All cells spend most of their time in *interphase*. While closely associated with mitosis and meiosis (and often mistakenly listed as the first step), this resting phase precedes the division process. In interphase, cells go about their daily business and slowly proceed through steps that prepare them to split in two — namely growth and copying all the DNA. (See Chapter 3 for more on this and other cell terminology). Meiosis, which you can see in Figure 16-3, is described in a series of stages as follows:

1. **Prophase I:** Structures disappear from the nucleus, including the nuclear membrane, nucleoplasm, and nucleoli. The cell's centrosomes, each containing two *centrioles,* move to opposite ends of the cell and form poles. Structures begin to appear in the nuclear region, including *spindles* (protein filaments that extend between the poles) and *asters,* or *astral rays* (protein filaments that extend from the poles into the cytoplasm). The *chromatin* (long strands of DNA) condenses to form *chromatids* which then bind together to form the familiar X-shape of the *chromosome,* or *sister chromatids.* *Homologous* chromosomes that contain the same genetic information pair up and go into *synapsis,* twisting around each other to form a *tetrad* of four chromatids. These tetrads begin to migrate toward the *equatorial plane* (an imaginary line between the poles).

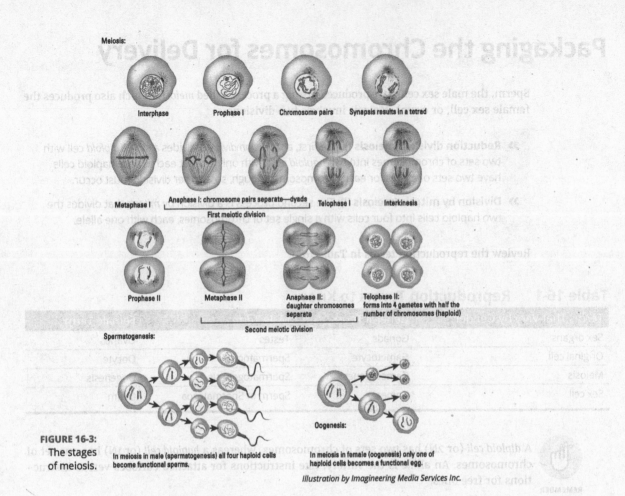

Meiosis:

Interphase | Prophase I | Chromosome pairs | Synapsis results in a tetrad

Metaphase I | Anaphase I: chromosome pairs separate—dyads | Telophase I | Interkinesis

First meiotic division

Prophase II | Metaphase II | Anaphase II: daughter chromosomes separate | Telophase II: forms into 4 gametes with half the number of chromosomes (haploid)

Second meiotic division

Spermatogenesis:

Oogenesis:

FIGURE 16-3:
The stages of meiosis.

In meiosis in male (spermatogenesis) all four haploid cells become functional sperms.

In meiosis in female (oogenesis) only one of haploid cells becomes a functional egg.

Illustration by Imagineering Media Services Inc.

2. **Metaphase I:** The tetrads align on the equatorial plane, attaching to the spindles by the centromere.

3. **Anaphase I:** Homologous chromosomes separate into dyads by moving along the spindles to opposite poles. In late anaphase, a slight furrowing is apparent in the cytoplasm, initiating *cytokinesis* (the division of the cytoplasm). Each pole receives one chromosome (2 identical chromatids).

4. **Telophase I:** The homologous chromosomes are now at opposite poles. Spindle and aster structures disappear, and the nucleus begins to rebuild in each newly forming cell. Chromosomes do not unwind here, they remain as sister chromatids, attached with the *centromere*. The furrowing seen in anaphase I continues to deepen as cytokinesis continues.

5. **Interkinesis:** The cytoplasm fully separates creating two haploid daughter cells. Though each contain one of every chromosome, they are not genetically identical as they can contain different alleles. In the male, the cytoplasm divides equally between the two cells. In the female, cytoplasmic division is unequal creating one cell and one polar body.

6. **Prophase II:** The cells enter the second phase of meiosis. Once again, structures disappear from the nuclei and poles appear at the ends. Spindles and asters appear in the nuclear region. Chromatin is already condensed into chromatids attached by the centromere, and they begin to migrate toward the equatorial plane.

7. **Metaphase II:** The chromatids align on the equatorial plane and the spindles attach to the centromere.

8. **Anaphase II:** The centromere splits and the chromatids separate, moving along the spindles to the poles. A slight furrowing appears in the cytoplasm.

9. **Telophase II:** With the chromatids at the poles, spindles and asters disappear while new nuclear structures appear. The chromatids uncoil, returning to their chromatin form. Cytoplasmic division continues via cytokinesis and each haploid cell divides, forming four cells each with one copy of each DNA strand.

At the end of this process, the male has four haploid sperm of equal size. As the sperm mature further, a *flagellum* (tail) develops. The female, on the other hand, has produced one large cell, the secondary oocyte, and three small cells called *polar bodies*; all four structures contain just one set of chromosomes. The polar bodies eventually disintegrate and the secondary oocyte becomes the functional cell. When fertilized by the sperm, the resulting *zygote* (fertilized egg) is diploid, containing two sets of chromosomes.

WARNING The terminology in mitosis/meiosis is especially confusing due to multiple uses of the word chromosome. We use the term to refer to our 23 pieces of DNA that contain all our genes. We, however, need two copies of every gene in our cells to function. So we say we have 46 chromosomes total — two copies (with different gene varieties) of each of the 23 pieces. These are not the x-shaped chromosomes we are familiar with though, hence the confusion. Those are created when we copy the 46 pieces of DNA to prepare for division and link them together with their identical match. Our "chromosomes" are really the propeller-shaped chromatids — one complete copy of the piece of DNA. So when we say our cells have two copies of each chromosome, we have two propellers, not two Xs (or four propellers joined in pairs). To further complicate the matter, when cells are not actively preparing for division (and most of the time they aren't), we don't see the propellers either. Chromatids are condensed DNA strands, formed by taking chromatin (the term for the unraveled strand of DNA) and wrapping it around proteins to make the DNA easier to move around. Without this condensation, our cells would be trying to equally split a giant plate of spaghetti which is hard enough as is, but add in the fact that which pieces go to which plate matters, it would be an impossible task without organizing it first.

Think you've conquered this process? Find out by tackling these practice questions:

EXAMPLE

Q. The metaphase II stage in meiosis involves

 a. the slipping of the centromere along the chromosome.

 b. the alignment of the chromosomes on the equatorial plane.

 c. the contraction of the chromosomes.

 d. the disappearance of the nuclear membrane.

A. The correct answer is the alignment of the chromosomes on the equatorial plane. Think "metaphase, middle."

---

17 A man has 46 chromosomes in a spermatocyte. How many chromosomes are in each sperm?

 a. 23 pairs

 b. 23

 c. 184

 d. 46

 e. 46 pairs

18 Synapsis, or side-by-side pairing, of homologous chromosomes

 a. occurs in mitosis.

 b. completes fertilization.

 c. occurs in meiosis.

 d. signifies the end of meiosis I.

 e. signifies the end of prophase of the second meiotic division.

19 Anaphase I of meiosis is characterized by which of the following?

   a. Synapsed chromosomes move away from the poles.

   b. DNA duplicates itself and condenses into chromosomes.

   c. Synapsis of homologous chromosomes occurs.

   d. Homologous chromosomes separate and move poleward with centromeres intact.

   e. Chromosomes are split at the centromere and pulled away from the poles.

20 During oogenesis, the three nonfunctional cells produced are called

   a. cross-over gametes.

   b. spermatozoa.

   c. polar bodies.

   d. oogonia.

   e. somatic cells.

21 Crossing over is a key process in increasing our genetic variability. It occurs when two chromatids (of the same chromosome but with different alleles) basically swap ends, mixing up the varieties of the genes it's carrying. During which phase is this most likely to occur?

   a. Prophase II

   b. Metaphase II

   c. Anaphase I

   d. Anaphase II

   e. Prophase I

22 Complete the following worksheet on the stages of meiosis. Draw the stages of meiosis and describe the changes in each stage.

1.

Late Prophase I

2.

Metaphase I

3.

Anaphase I

4.

Telophase I

5.

Interkinesis

6.

Prophase II

7.

Metaphase II

8.

Anaphase II

9.

Telophase II

10.

Sperm

11.

Ovum and Polar Bodies

Answers to Questions on the Male Reproductive System

The following are answers to the practice questions presented in this chapter.

(1) Why does the left side of the scrotum tend to hang lower than the right? **c. To accommodate a longer spermatic cord.** The key words here are "tend to." It's not true of all men, but in general the left testicle hangs lower than the right because it has a longer cord.

(2) The **c. dartos tunic** is the source of the action when the scrotum adjusts to surrounding temperatures. This is the inner smooth muscle layer of the scrotum.

(3-8) Sperm-forming cells, or **3. spermatogonia,** divide throughout the reproductive lifetime of the male by a process called **4. mitosis.** But after puberty, these cells begin to produce two types of **5. daughter** cells. Type A cells remain in the wall of the **6. seminiferous tubules,** producing more of themselves. Type B cells produce **7. four sperm** through a process called **8. meiosis.**

(9) Where is testosterone produced? **b. Leydig (interstitial) cells.** The word *interstitial* can be translated as "placed between," which is where Leydig cells are.

(10) Select the correct sequence for the movement of sperm: **e. Seminiferous tubules → tubuli recti → rete testis → epididymis → ductus deferens → ejaculatory duct → urethra.** The sperm develop in the coiled tubules, move through the straighter tubes (tubuli recti), continue across the network of the testis (rete testis), through the efferent ducts, into the epididymis (remember the really long tube), and past the ductus (or vas) deferens and the ejaculatory duct into the urethra.

(11) As the sperm moves through the reproductive tract, which of the following does not add a secretion to it? **a. Leydig (interstitial) cells.** Interstitial cells secrete testosterone, the primary male hormone.

(12) Sperm storage and development takes place in the convoluted tube called the **e. epididymis.** The other answer options don't come into play until it's time to release semen.

(13) After all the proper glands have secreted fluids to nourish and protect the departing sperm, the substance ejaculated is called **b. semen.**

(14) An average ejaculation will contain about **b. 400 to 500 million** sperm. Keep in mind that sperm are microscopically small, so quite a few can fit in a tiny amount of semen.

(15) During sexual arousal, the penis becomes hard and erect. What is happening to cause that? **c. Blood vessels are dilating to allow free flow of blood into the penis, while vascular shunts constrict drainage and cause blood pressure to rise in the corpora cavernosa.**

(16) Explain the difference between emission and ejaculation: **While both are under sympathetic control and seem to happen simultaneously, there is an important distinction. Erection is triggered via a parasympathetic pathway so it's important to note that emission doesn't begin until repeated stimulation initiates it. A simple way to think about it is emission is preparation and ejaculation is release.**

(17) A man has 46 chromosomes in a spermatocyte. How many chromosomes are in each sperm? **b. 23.** Because the spermatocyte is the original cell that undergoes division, and because a human has a total of 46 chromosomes only *after* sperm meets egg, you must divide the number 46 in half.

18. Synapsis, or side-by-side pairing, of homologous chromosomes **c. occurs in meiosis.** Specifically, synapsis occurs during prophase I, which is the first stage of meiosis.

19. Anaphase I of meiosis is characterized by which of the following? **d. Homologous chromosomes separate and move poleward with centromeres intact.** Think anaphase, apart. Then you just have to take note of whether it's meiosis I and II to determine what is being pulled apart.

20. During oogenesis, the three nonfunctional cells produced are called **c. polar bodies.** They eventually disintegrate.

21. Crossing over is a key process in increasing our genetic variability. It occurs when two chromatids (of the same chromosome but with different alleles) basically swap ends, mixing up the varieties of the genes it's carrying. During which phase is this most likely to occur? **e. Prophase I**

22. Following is a summary of what should appear in your drawings and descriptions of the stages of meiosis. For further reference, check out Figure 16-3.

In the drawing for late prophase I, at least two pairs of homologous chromosomes should be grouped into tetrads. (In truth, there are 23 pairs, but simplified illustrations tend to show just two.) The description for prophase I should include reference to the tetrad formation. The drawing for metaphase I should show the equatorial plane (a center horizontal line) with the tetrads aligned along it; an "x" on each side. The illustration also should show spindles radiating from each pole, with the tetrads attached to them by their centromeres. The description should include reference to the equatorial plane, the poles, and the spindles.

The drawing for anaphase I should show the tetrads moving to the top and bottom of the cell along the spindles and the cytoplasm slowly beginning to divide. In telophase I, the division becomes more pronounced and two new nuclei form. The chromosomes remain in the "x" shape. As the process enters interkinesis, the cytoplasm pinches off into two cells.

During prophase II, which also is the start of the second meiotic division, the chromosomes migrate toward a new equatorial plane. The drawing of metaphase II should show all chromosomes aligned on the equatorial plane with the centromere along the (invisible) center line. For anaphase II, you should show the chromosomes pulling apart into chromatids and moving toward the poles. In the final stage, telophase II, you should draw new nuclei forming around the chromatin as it uncoils at this point.

Chapter 17

Carrying Life: The Female Reproductive System

Men may have quite a few hard-working parts in their reproductive systems (see Chapter 16), but women are the ones truly responsible for survival of the species (biologically speaking, anyway). The female body prepares for reproduction every month for most of a woman's adult life, producing a secondary oocyte and then measuring out delicate levels of hormones to prepare for nurturing a developing embryo. When a fertilized ovum fails to show up, the body hits the biological reset button and sloughs off the uterine lining before building it up all over again for next month's reproductive roulette. But that's nothing compared to what the female body does when a fertilized egg actually settles in for a nine-month stay. Strap yourselves in for a tour of the incredible female baby-making machinery — practice questions included.

Identifying the Female Reproductive Parts and Their Functions

Before we dive in to major reproductive function, let's take a brief tour of the female reproductive system's parts.

External genitalia

Unlike in the male, most of a female's reproductive organs are internal. However, there are external ones that play a role. Together, the external structures are referred to as the *vulva*. These include:

» **Mons pubis:** Fatty tissue that lies atop the pubic symphysis, secretes pheromones, and grows hair to trap particles, preventing them from entering the reproductive tract

» **Labia majora:** The outer folds of tissue that protect the area, contain sweat and sebaceous (oil) glands to aid in lubrication, homologous to the scrotum

» **Labia minora:** Surround the openings to the urethra and vagina, are highly vascularized and become engorged during arousal to increase sensitivity (to pleasurable sensation)

» **Clitoris:** A small mound of erectile tissue on the anterior end of where there the labia minora meet, it is the most sensitive area of the female genitalia — repeated stimulation can lead to orgasm; homologous to the penis

» **Bartholin glands:** Located at the opening of the vagina, secretes a thick fluid for lubrication during intercourse

Internal genitalia

The remaining organs are all found internally in the reproductive tract. These include:

» **Vagina:** Opens to the external genitalia, receives penis for intercourse; serves as *birth canal* for childbirth

» **Cervix:** A ring of muscular and connective tissue at the end of the vagina, guarding the uterus, produces a thick mucus to trap particles and bacteria except before ovulation where it thins to allow passage for sperm

» **Uterus:** Pear-shaped organ for housing a developing fetus, the lining, or *endometrium*, is triggered by estrogen to add bulk early in a woman's cycle to prepare it for the embryo; if fertilization does not occur, the lining is shed during *menstruation* and the process starts over

» **Uterine tubes (also known as *Fallopian tubes* or *oviducts*):** Attached at the top right and left sides of the uterus, widen at the ovary ends into the fan-shaped *infundibulum*, site of fertilization

» **Ovaries:** The almond-shaped gonads, produce estrogen and progesterone, develop eggs

The term "estrogen" actually refers to a group of hormones primarily produced in the ovaries. These are responsible for developing the reproductive organs as well as providing the female secondary sex characteristics (as testosterone does in males). The primary estrogen hormone is *estradiol*, so when estrogen is used generically, this is the hormone being referred to.

Gamete production

First and foremost, in the female reproductive repertoire are the two *ovaries*, which usually take a turn every other month to produce a single *ovum* or *secondary oocyte*. A sheet of connective tissue called the broad ligament holds the ovaries in place. Each ovary is surrounded by a dense fibrous connective tissue called the *tunica albuginea* (literally "white covering"); yes, that's the same name as the tissue surrounding the testes. In fact, the ovaries in a female and the testes in a male are *homologous*, meaning that they share similar origins. The ovaries themselves are arranged into an outer layer (the cortex) and an inner layer (the medulla). The cortex houses the *primordial follicles*, which will eventually mature into ova. The medulla is loosely packed with blood and lymphatic vessels and nerves.

Ovarian cycle

During fetal development, cells in the ovaries called *oogonia* begin meiosis, creating about two million *primary oocytes* suspended in prophase I. By puberty, though, about three-quarters of them have died. At the onset of puberty, the primary oocytes resume meiosis, splitting unevenly into a *polar body* and the *secondary oocyte*. These are wrapped in a single layer of epithelial tissue creating the *primary follicle*. They continue meiosis and the polar body (which is just a nucleus in a cell membrane) completes the process creating two polar bodies which will just degenerate. The secondary oocyte pauses meiosis again, this time at metaphase II. It will not complete meiosis until fertilization has been initiated — finally creating the actual ovum and a polar body that will degenerate.

At puberty and approximately once each month until menopause (which we cover in the later section "Growing, Changing, and Aging"), the *ovarian cycle* occurs as follows:

1. The anterior pituitary gland secretes *follicle-stimulating hormone*, or *FSH*, which prompt about 1,000 of the primordial follicles to resume meiosis.

    Usually, one follicle outgrows the rest, maturing into a *Graafian follicle.* Non-identical, or *fraternal*, twins, triplets, or even more embryos result if more than one follicle matures and more than one secondary oocyte is released (and fertilized).

2. Follicular cells divide and surround the oocyte with several layers of cells called the *cumulus oophorus*, creating the *secondary follicle*.

3. As maturation continues, the cell layers separate leaving a fluid-filled space between them. The cells in contact with the oocyte form the *zona pelludica*. The others secrete estrogen to signal the uterus to prepare by thickening the endometrium.

4. As blood levels of estrogen begin to rise, the pituitary stops releasing FSH and begins releasing *luteinizing hormone*, or *LH*, which prompts the Graafian follicle now at the surface of the ovary to rupture, triggering *ovulation* — the release of the secondary oocyte, more commonly referred to as an egg cell.

5. The oocyte, wrapped in the zona pellucida, takes some follicle cells with it when released, which form the *corona radiata*. They serve to provide nourishment for the journey. Since the uterine tube is not directly connected to the ovary, the egg is released into the pelvic cavity. However, the end of the uterine tube, the infundibulum, is crowned in finger-like projections called *fimbriae*. Cells of this tissue are ciliated, which wave at ovulation to usher the egg into the tube.

6. It takes approximately three days for the egg to travel through the uterine tube to the uterus, regardless of whether or not it is fertilized. The egg is only viable, however, for about 24 hours.

Meanwhile, back at the ovary, a clot has formed inside the ruptured follicle, and the remaining follicle cells form the *corpus luteum* (literally "yellow body"). This new but temporary endocrine gland secretes *progesterone*, a hormone that signals the uterine lining to prepare for possible implantation of a fertilized egg; inhibits the maturing of follicles, ovulation, and the production of estrogen to prevent menstruation; and stimulates further growth in the mammary glands (which is why some women get sore breasts a few days before their periods begin). If pregnancy occurs, the placenta also will release progesterone to prevent menstruation throughout the pregnancy.

If the egg is not fertilized, the corpus luteum dissolves after 10 to 14 days to be replaced by scar tissue called the *corpus albicans*. If pregnancy does occur, the corpus luteum remains and grows for about six months before disintegrating. Only about 400 of a woman's primordial follicles ever develop into secondary oocytes for the trip to the uterus. Figure 17-1 shows what happens inside an ovary.

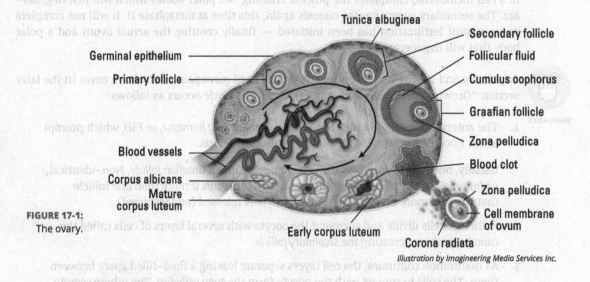

FIGURE 17-1:
The ovary.

Illustration by Imagineering Media Services Inc.

Uterine cycle

Occurring alongside the ovarian cycle is the *uterine cycle*, which is directly influenced by the hormones made in the ovaries. Day one of the cycle occurs at the onset of menstruation, when a woman "starts her period" so to speak. Because the egg was not fertilized, the uterus must hit reset (the *menstrual phase*) and shed the endometrium; otherwise it would not be hospitable for the new embryo when it arrives following the next ovarian cycle. After the ovaries start releasing estrogen, the uterus begins its *proliferation phase* by creating more connective tissue, thickening the endometrium. Following ovulation and the release of progesterone, the uterus enters the *secretory phase* and increases blood flow to the endometrium in order to provide nutrients for the embryo. (For more on this process, flip ahead to the "Making Babies: An Introduction to Embryology" section.)

The middle layer of the uterine wall, the *myometrium*, is comprised of smooth muscle. It contracts during menstruation to slough off the inner lining. If pregnancy is achieved, progesterone blocks any uterine contraction until the onset of labor. At that point, the uterus begins contracting with increasing frequency and force to push the fetus through the birth canal.

Find out how familiar you are with the female reproductive structures and their functions:

---

1 True or False: The medulla of the ovary is the site of follicular development.

2-11 Fill in the blanks to complete the paragraph.

The 2. _____ in a developing female fetus begin meiosis. Upwards of 2 million

3. _____ are produced but they are suspended in 4._____.

By the onset of 5._____ only about 300,000–400,000 remain. Each month, about 1,000 primary follicles are triggered to resume meiosis by 6._____. One usually outgrows the rest, forming the

7._____. Meiosis is again suspended, this time at

8._____. Release of the secondary oocyte, or 9._____, is triggered by the release of 10._____ from the anterior pituitary. Meiosis II is not completed until 11._____ has begun.

12–16 Match the term to its description.

a. Fingerlike projections at the end of a uterine tube

b. Protective layer directly surrounding the secondary oocyte

c. Granulosa cells forming the outer layer of the Graafian follicle

d. Layer of smooth muscle in uterine wall

e. Inner lining of the uterus

12 _____ Corona radiata

13 _____ Endometrium

14 _____ Fimbriae

15 _____ Zona pelludica

16 _____ Myometrium

17 Which one of the following is NOT a function of estradiol?

a. Preparing the endometrium

b. Supporting development of the secondary oocyte

c. Supporting development of the corpus luteum

d. Triggering development of the reproductive organs

e. Preventing secretion of FSH from the pituitary

18–33 Use the terms that follow to identify the anatomy of the female reproductive system shown in Figure 17–2.

18. _____ _____

19. _____ _____

20. _____ _____

21. _____ _____

22. _____ _____

23. _____ _____

24. _____ _____

25. _____ _____

26. _____ _____

27. _____ _____

28. _____ _____

29. _____ _____

30. _____ _____

31. _____ _____

32. _____ _____

33. _____ _____

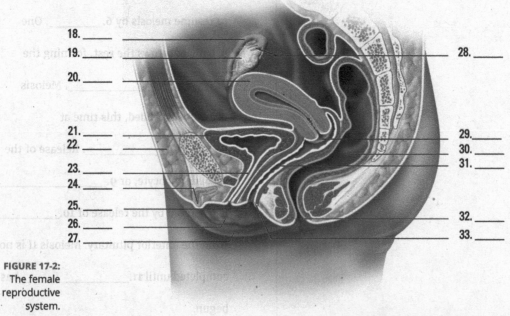

FIGURE 17-2:
The female reproductive system.

Illustration by Imagineering Media Services Inc.

a. Cervix

b. Fimbriae

c. Urinary bladder

d. Clitoris

e. Labium major

f. Ovary

g. Rectum

h. Bartholin's glands

i. Vaginal orifice

j. Anus

k. Mons pubis

l. Uterus

m. Labium minor

n. Uterine (Fallopian) tube

o. Vagina

p. Urethra

---

34 Why does the corpus luteum produce progesterone?

a. To stimulate development of the cumulus oophorus

b. To trigger ovulation

c. To prepare a woman's uterus for pregnancy and prevent menstruation

d. To trigger the resumption of meiosis

e. To prepare the infundibulum to fragment into fimbriae

35 True or False: Once a sperm penetrates the cell membrane of the ovum, the zona pelludica immediately hardens.

Making Eggs: A Mite More Meiosis

We cover the details of meiosis in Chapter 16, but in this section we explore how meiosis contributes to creating a *zygote* (fertilized egg) and its chromosomal makeup. Normal human diploid, or 2N, cells contain 23 pairs of homologous chromosomes for a total of 46 chromosomes each. One chromosome from each pair comes from the individual's mother, and the other comes from the father. Each homologous pair contains the same type of genetic information, but the *allele* (variety) of this genetic information may differ from one chromosome of the pair to the homologous one. For example, one chromosome from the mother may carry the genetic coding for blue eyes, whereas the homologous chromosome from the father may code for brown eyes.

REMEMBER

Although all the body's cells have 46 chromosomes — including the cells that eventually mature into the ovum and the sperm — each gamete (either ovum or sperm) must offer only half that number if fertilization is to succeed. That's where meiosis steps in. As you see in Figure 17-3, the number of chromosomes is cut in half during the first meiotic division, producing gametes that are haploid, or 1N. They, however, still contain two alleles for each gene so another division must occur. The second meiotic division produces four haploid sperms in the male but only one functional haploid secondary oocyte in the female. When fertilization occurs, the new zygote will contain 23 pairs of homologous chromosomes for a total of 46 chromosomes. The zygote then proceeds through mitosis to produce the body's cells, distributing copies of all 46 chromosomes to each new cell.

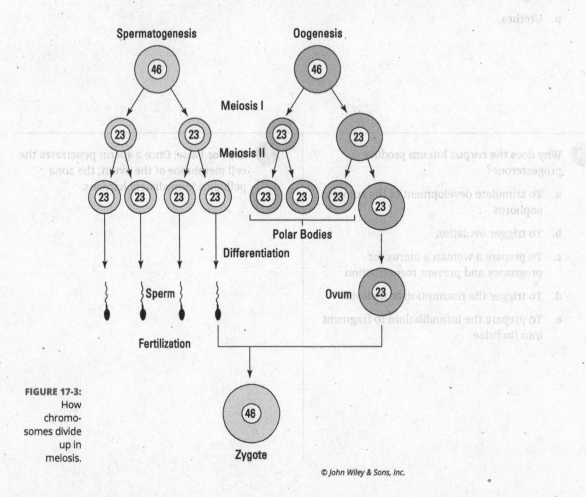

FIGURE 17-3: How chromosomes divide up in meiosis.

© John Wiley & Sons, Inc.

Think you've conquered this process? Find out by answering the following practice questions:

EXAMPLE

Q. How many chromosomes are in most of the cells in your body?

a. 46

b. 23

c. 92

A. The correct answer is 46. That's the usual human complement.

36-44 Fill in the blanks to complete the following sentences:

Meiosis produces sperm and ova, which when combined make a **36.**_____ (fertilized egg) with its full complement of chromosomes. Normally, humans have **37.**_____, or 2N, cells containing 23 pairs of homologous chromosomes for a total of 46 chromosomes each. A pair of chromosomes containing the same type of genetic information are **38.**_____ chromosomes though they may have different **39.**_____ or varieties of a gene. Ova and sperm are called sex cells or **40.**_____. The number of chromosomes is cut in half during the first meiotic division, producing gametes that are **41.**_____, or 1N. The second meiotic division produces **42.**_____ sperm in the male but only **43.**_____ secondary oocyte in the female. The zygote then proceeds through **44.**_____ to produce the body's cells.

Making Babies: An Introduction to Embryology

Fertilization (the penetration of a secondary oocyte by a sperm) must occur in the uterine tube within 24 hours of ovulation or the secondary oocyte degenerates. But that doesn't mean that sexual intercourse outside that time frame can't lead to pregnancy. Research indicates that spermatozoa can survive up to seven days inside the female reproductive tract. If a sperm is still motile — that is, if it's still whipping its flagellum tail — when an ovum comes down the tube, it will do what it was made to do and penetrate the ovum's membrane.

The sperm releases enzymes that allow it to digest its way into the secondary oocyte, leaving its flagellum behind. With the penetration of the sperm, meiosis II is completed, resulting in an ovum and a second polar body. When the *pronuclei* of the ovum and the sperm unite, presto! You've got a zygote (fertilized egg). Over the next three to five days, the zygote moves through the uterine tube to the uterus, undergoing *cleavage* (mitotic cell division) along the way: Two cells become four smaller cells, four cells become eight smaller cells, and then those eight cells become a solid 16-cell ball called a *morula*. After five days of cleavage, the cells form a hollow ball called a *blastula*, or *blastocyst*. The inner hollow region is called the *blastocoele*; the outer-layer cells are called the *trophoblast*; and the inner mass of cells is the *embryoblast*. Figure 17-4 illustrates this process of development.

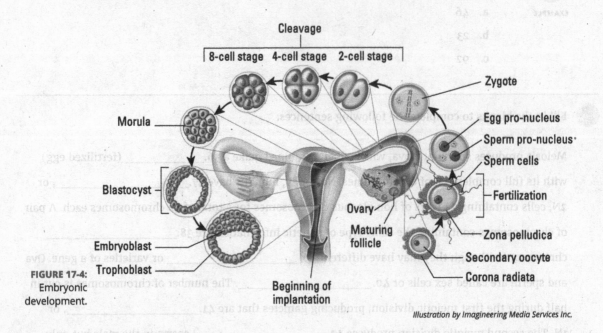

FIGURE 17-4: Embryonic development.

Illustration by Imagineering Media Services Inc.

Within three days of its arrival in the uterus (generally within a week of fertilization), the blastocyst implants in the endometrium, and some of the blastocyst's cells — called *totipotent embryonic stem cells* — organize into an inner cell mass also known as the *embryonic disk*. (Refer to Chapter 3 for more on stem cells.) Over time, the embryonic disk *differentiates* into the tissues of the developing embryo. The outer cells of the disk form the *amnion* and its resulting *amniotic cavity*, and the inner ones form two primitive *germ layers*. The layer nearest the amniotic cavity forms the *ectoderm*, while the others form the *endoderm*. Between the two layers, cells differentiate to form a third layer, the *mesoderm*. Cells of the three germ layers develop into the body parts:

>> **Ectoderm:** Skin, nervous tissue

>> **Mesoderm:** Bones, cartilage, connective tissue, muscle

>> **Endoderm:** Respiratory and digestive tract organs, glands

TIP

To keep these terms straight, remember that *endo*– means "inside or within," *ecto*– means "outer or external," and *meso*– means "middle."

After three weeks of development, the heart begins beating. In the fourth week of development, the embryonic disk forms an elongated structure that attaches to the developing placenta by a connecting stalk. A head and jaws form while primitive buds sprout; the buds will develop into arms and legs. During the fifth through seventh weeks, the head grows rapidly and a face begins to form (eyes, nose, and a mouth). Fingers and toes grow at the ends of the elongating limb buds. All internal organs have started to form. After eight weeks of development, the embryo begins to have a more human appearance and is referred to as a *fetus*.

The outer cells of the blastocyst, together with the endometrium of the uterus, form the *placenta*, a new internal organ that exists only during pregnancy. The placenta attaches the fetus to the uterine wall, exchanges gases and waste between the maternal and fetal bloodstreams, and secretes hormones to sustain the pregnancy.

Now that you've refreshed your memory a bit about how babies are made (beyond the birds and bees talks), try the following practice questions:

45-53 Mark the statement with a T if it's true or an F if it's false:

45 _____ The mesoderm will form nervous tissue and skin.

46 _____ The embryonic stage is complete at the end of the eighth week.

47 _____ Outer cells of the embryonic disk form the germ layers.

48 _____ During the fifth through seventh weeks, the arm and leg buds elongate, and fingers and toes begin to form.

49 _____ Cleavage is successive mitotic divisions of the embryonic cells into smaller and smaller cells.

50 _____ As the zygote moves through the Fallopian tube, it undergoes meiosis.

51 _____ The placenta exchanges gases and waste between the maternal blood and the fetal blood.

52 _____ After five days of cleavage, the cells form into a hollow ball called the morula.

53 _____ Sexual intercourse five days before ovulation cannot lead to pregnancy.

Growing from Fetus to Baby

Pregnancy is divided into three periods called *trimesters* (although many new parents bemoan a postnatal fourth trimester until the baby sleeps through the night). The first 12 weeks of development mark the first trimester, during which *organogenesis* (organ formation) is established. During the second trimester (typically considered to be weeks 13 to 27), all fetal systems continue to develop and rapid growth triples the fetus's length. By the third trimester (typically considered to be weeks 28 to 40), all organ systems are functional and the fetus usually is considered *viable* (capable of surviving outside the womb) even if it's born prematurely. The overall growth rate slows in the third trimester, but the fetus gains weight rapidly.

Milestones in fetal development can be marked monthly.

REMEMBER

>> At the **end of the second month** (when the terminology changes from "embryo" to "fetus"), the head remains overly large compared to the developing body, and the limbs are still short. All major regions of the brain have formed.

>> During the **third month**, body growth accelerates and head growth slows. The arms reach the length they will maintain during fetal development. The bones begin to ossify, and all body systems have begun to form. The circulatory (cardiovascular) system supplies blood to all the developing extremities, and even the lungs begin to practice "breathing" amniotic fluid. By the end of the third month, the external genitalia are visible in the male (ultrasound technicians call this a "turtle sign"). The fetus is a bit less than 4 inches long and weighs about 1 ounce.

>> The body grows rapidly during the **fourth month** as legs lengthen, and the skeleton continues to ossify as joints begin to form. The face looks more human. The fetus is about 7 inches long and weighs 4 ounces.

>> Growth slows during the **fifth month,** and the legs reach their final fetal proportions. Skeletal muscles become so active that the mother can feel fetal movement. Hair grows on the head, and *lanugo*, a profusion of fine, soft hair, covers the skin. The fetus is about 12 inches long and weighs ½ to 1 pound.

>> The fetus gains weight during the **sixth month,** and eyebrows and eyelashes form. The skin is wrinkled, translucent, and reddish because of dermal blood vessels. The fetus is between 11 and 14 inches long and weighs a bit less than 1½ pounds.

>> During the **seventh month,** skin becomes smoother as the fetus gains subcutaneous fat tissue. The eyelids, which are fused during the sixth month, open. Usually, the fetus turns to an upside-down position. It's between 13 and 17 inches long and weighs from 2½ to 3 pounds.

>> During the **eighth month,** subcutaneous fat increases, and the fetus shows more baby-like proportions. The testes of a male fetus descend into the scrotum. The fetus is now 16 to 18 inches long and has grown to just under 5 pounds.

>> During the **ninth month,** the fetus plumps up considerably with additional subcutaneous fat. Much of the lanugo is shed, and fingernails extend all the way to the tips of the fingers. The average newborn at the end of the ninth month is 20 inches long and weighs about 7½ pounds.

The following practice questions deal with the development of the fetus during its 40 weeks in the womb:

---

54 What needs to happen before a fetus can be considered viable?

   a. The heart rate must be easily detected.

   b. All organ systems must be functional.

   c. Subcutaneous fat must be built up to at least an inch.

   d. The brain is fully formed.

   e. The fetus must have turned into an upside-down position.

55 Describe one new fetal development for each month:

3rd month: _____

4th month: _____

5th month: _____

6th month: _____

7th month: _____

8th month: _____

9th month: _____

Parturition

Near the end of the fetus' stay in the womb, the woman's body begins preparations for *parturition*, or childbirth. Progesterone steadily drops as the big day nears leading to mild contractions that most women don't even feel. The cervix begins to thin (called *effacement*) and dilate to allow the fetus into the birth canal (vagina). But the trigger to set off the cascade of hormones that maintain the process of *labor* still remains a mystery. It is widely believed that it is initiated by the fetus. Regardless of the trigger, parturition is a classic example of a *positive feedback loop*.

Prostaglandins cause early uterine contractions, which leads to stretch (in the non-contracting parts) of the uterus. This feeds back to the hypothalamus, causing it to release *oxytocin* from the posterior pituitary. This triggers uterine contractions, which causes more stretch, which causes more oxytocin and more contractions, and so on, until the baby is delivered.

Labor progresses through three stages, the first of which is the longest. Stage one is split into three phases (that any woman who has experienced them would prefer to forget):

>> **Phase 1 — latent labor:** Contractions every 5–20 minutes, cervix dilates up to 5–6 cm, typically lasts from 6–12 hours

>> **Phase 2 — active labor:** Increasingly stronger contractions every 3–5 minutes that last about 60 seconds, cervix dilates up to 8 cm, typically lasts from 3–5 hours

>> **Phase 3 — transition:** Intense contractions every 2–3 minutes for 60–90 seconds, cervix fully dilates, typically lasts from 30–60 minutes — most women say this is the most painful part of the entire childbirth process

Transition prepares the body for the second stage of labor as contractions shift from an all-around squeeze to a downward push. A woman feels the urge to push with each contraction, which slow down to every 3–5 minutes and last for 60–75 seconds. It takes, on average, 20 minutes to 2 hours to complete this stage by delivering the baby. After a brief reprieve, weaker contractions resume to deliver the placenta. This final stage of labor generally take less than 30 minutes.

The hours following parturition are critical for the bond between mother and child. *Prolactin* begins triggering milk production in the mammary glands, and the nursing infant stimulates the release of oxytocin to release the milk, called *let down*. Oxytocin is also known as the "love hormone"; it's the neurotransmitter that provides the love sensation we feel for family.

Are you feeling the love? Let's find out by answering some questions:

56 Which hormones play an active role in the progression of labor?

      I.   Progesterone

     II.   Prostaglandins

    III.   Oxytocin

a. I only

b. III only

c. I and II

d. II and III

e. I, II, and III

57 Which stage of labor usually lasts the longest?

a. stage 1, phase 1

b. stage 1, phase 2

c. stage 1, phase 3

d. stage 2

e. stage 3

58 Explain how childbirth is an example of a positive feedback loop.

_____

_____

_____

Growing, Changing, and Aging

From birth to 4 weeks of age, the newborn is called a *neonate*. Faced with survival after its physical separation from the mother, a neonate must abruptly begin to process food, excrete waste, obtain oxygen, and make circulatory adjustments.

From 4 weeks to 2 years of age, the baby is called an *infant*. Growth during this period is explosive under the stimulation of *growth hormone* from the pituitary gland, adrenal steroids, and thyroid hormones. The infant's *deciduous* teeth, also called *baby* or *milk teeth*, begin to form and erupt through the gums. The nervous system advances, making coordinated activities possible. The baby begins to develop language skills.

From 2 years to puberty, you're looking at a *child*. Influenced by growth hormones, growth continues its rapid pace as deciduous teeth are replaced by permanent teeth. Muscle coordination, language skills, and intellectual skills also develop rapidly.

From puberty, which starts between the ages of 11 and 14, to adulthood, the child is called an *adolescent*. Growth occurs in spurts. Girls achieve their maximum growth rate between the ages of 10 and 13, whereas boys experience their fastest growth between the ages of 12 and 15. Primary and secondary sex characteristics begin to appear. Growth terminates when the epiphyseal plates of the long bones ossify sometime between the ages of 18 and 21 (see Chapter 7 for an introduction to the bones of the body). Motor skills and intellectual abilities continue to develop, and psychological changes occur as adulthood approaches.

The young adult stage covers roughly 20 years, from the age of 20 to about 40. Physical development reaches its peak and adult responsibilities are assumed, often including a career, marriage, and a family. By age 25, brain development is complete, marked by the maturation of the *prefrontal cortex* — the part of the brain responsible for *executive functions* like planning, decision making, and problem solving. After about age 30, physical changes that indicate the onset of aging begin to occur.

From age 40 to about 65, physiological aging continues. Gray hair, diminished physical abilities, skin wrinkles, and other outward signs of aging are caused by the decreased activity of cells and the lack of their replacement. Most notably, collagen and elastin fibers break down and are not replaced.

Women go through *menopause,* the cessation of monthly cycles. While *menarche* (the onset of menstruation) may begin any time between 10 and 15 years of age, the female body's monthly reproductive cycle slows and stops entirely between the ages of 40 and 55. With the cessation of menses comes a decrease in size of the uterus, shortening of the vagina, shrinkage of the mammary glands, disappearance of Graafian follicles, and shrinkage of the ovaries. For about six years prior to menopause, many women experience a stage called *perimenopause* during which increasingly irregular hormone secretions can cause fluctuations in menstruation, a sensation called *hot flashes,* and mood irritability. At around the same age, males may experience *andropause* due to a drop in testosterone. However, this drop is gradual and often does not completely cease, as estrogen does in women. Some men feel fatigued, irritable, and depressed, and they experience lack of sex drive, though others maintain their drive as well as the reproductive functionality (production of sperm) their whole life.

REMEMBER

From age 65 until death is the period of *senescence,* the gradual decline in function we call aging. Individual adults can show widely varying patterns of aging in part because of differences in genetic background and physical activities. Signs of senescence include

» Loss of skin elasticity and accompanying sagging or wrinkling

» Weakened bones and decreasingly mobile joints

» Weakened muscles

» Impaired coordination, memory, and intellectual function

» Cardiovascular problems

» Reduced immune responses

» Decreased respiratory function caused by reduced lung elasticity

» Decreased peristalsis and muscle tone in the digestive and urinary tracts

Following are some practice questions on the aging cycle:

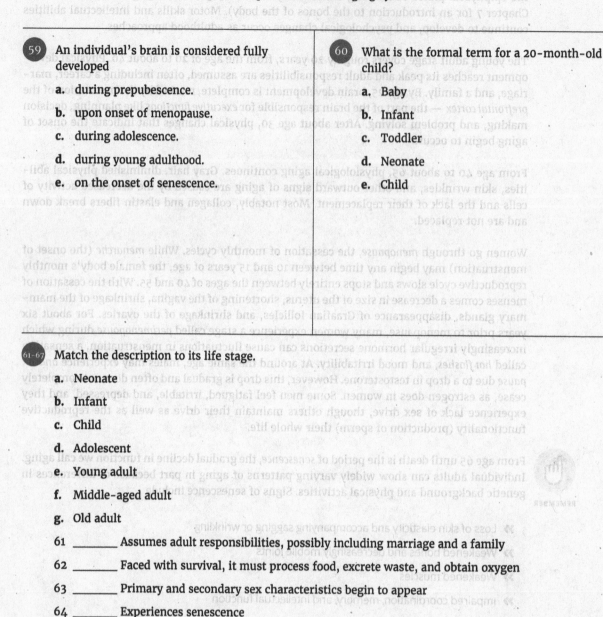

59 An individual's brain is considered fully developed
  a. during prepubescence.
  b. upon onset of menopause.
  c. during adolescence.
  d. during young adulthood.
  e. on the onset of senescence.

60 What is the formal term for a 20-month-old child?
  a. Baby
  b. Infant
  c. Toddler
  d. Neonate
  e. Child

61-67 Match the description to its life stage.
  a. Neonate
  b. Infant
  c. Child
  d. Adolescent
  e. Young adult
  f. Middle-aged adult
  g. Old adult

61 _____ Assumes adult responsibilities, possibly including marriage and a family

62 _____ Faced with survival, it must process food, excrete waste, and obtain oxygen

63 _____ Primary and secondary sex characteristics begin to appear

64 _____ Experiences senescence

65 _____ Deciduous teeth begin to form

66 _____ Women go through menopause

67 _____ From 2 years of age to puberty

Answers to Questions on the Female Reproductive System

The following are answers to the practice questions presented in this chapter.

1 True or False: The medulla of the ovary is the site of follicular development. **False.** This occurs in the outer layer, the cortex; otherwise how would the egg get out?

2-11 Fill in the blanks to complete the paragraph.

The **2. oogonia** in a developing female fetus begin meiosis. Upwards of 2 million **3. primordial follicles** are produced but they are suspended in **4. prophase I**. By the onset of **5. puberty** only about 300,000–400,000 remain. Each month, about 1,000 primary follicles are triggered to resume meiosis by **6. FSH**. One usually outgrows the rest forming the **7. Graafian follicle**. Meiosis is again suspended, this time at **8. Metaphase II**. Release of the secondary oocyte, or **9. ovulation**, is triggered by the release of **10. LH** from the anterior pituitary. Meiosis II is not completed until **11. fertilization** has begun.

12 Corona radiata: **c. Granulosa cells forming the outer layer of the Graafian follicle**

13 Endometrium: **e. Inner lining of the uterus**

14 Fimbriae: **a. Fingerlike projections at the end of a uterine tube**

15 Zona pelludica: **b. Protective layer directly surrounding the secondary oocyte**

16 Myometrium: **d. Layer of smooth muscle in the uterine wall**

17 Which one of the following is not a function of estradiol? **c. Supporting development of the corpus luteum.** By the time the corpus luteum starts to develop, estrogen already has begun to bow out of the reproductive equation, allowing progesterone to take the lead.

18-33 Following is how Figure 17-2, the female reproductive system, should be labeled.

18. n. Uterine (Fallopian) tube; 19. f. Ovary; 20. l. Uterus; 21. c. Urinary bladder; 22. k. Mons pubis; 23. p. Urethra; 24. d. Clitoris; 25. i. Vaginal orifice; 26. m. Labium minor; 27. e. Labium major; 28. b. Fimbrae; 29. a. Cervix; 30. g. Rectum; 31. o. Vagina; 32. j. Anus; 33. h. Bartholin's glands

34 Why does the corpus luteum produce progesterone? **c. To prepare a woman's uterus for pregnancy and prevent menstruation.** It stands to reason that if the corpus luteum forms after the Graafian follicle ruptures, the progesterone it's producing is signaling the uterus that a fertilized egg may be on its way.

35 True or False: Once a sperm penetrates the cell membrane of the ovum, the zona pelludica immediately hardens. **True:** This prevents another sperm from entering; 3 of each chromosome would likely mean the embryo would not develop.

36-44 Meiosis produces sperm and ova, which when combined make a **36. zygote** (fertilized egg) with its full complement of chromosomes. Normal humans have **37. diploid,** or 2N, cells containing 23 pairs of homologous chromosomes for a total of 46 chromosomes each. A pair of chromosomes containing the same type of genetic information are **38. homologous** chromosomes, though they may have different **39. alleles** or varieties of a gene. Ova and sperm are called sex cells or **40. gametes.** The number of chromosomes is cut in half during the first meiotic division, producing gametes that are **41. haploid,** or 1N. The second meiotic division produces **42. four haploid** sperm in the male but only **43. one functional haploid** secondary oocyte in the female. The zygote then proceeds through **44. mitosis** to produce the body's cells.

45 The mesoderm will form nervous tissue and skin. **False.**

46 The embryonic stage is complete at the end of the eighth week. **True.**

47 Outer cells of the embryonic disk form the germ layers. **False.** The inner cells do this; the outer cells form the amniotic sac.

48 During the fifth through seventh weeks, the arm and leg buds elongate, and fingers and toes begin to form. **True.**

49 Cleavage is successive mitotic divisions of the embryonic cells into smaller and smaller cells. **True.**

50 As the zygote moves through the uterine tube, it undergoes meiosis. **False.** Meiosis is completed during fertilization.

51 The placenta exchanges gases and waste between the maternal blood and the fetal blood. **True.**

52 After five days of cleavage, the cells form into a hollow ball called the morula. **False.** When the embryo hollows out it becomes the blastocyst.

53 Sexual intercourse five days before ovulation cannot lead to pregnancy. **False.** It certainly can happen — and does. As long as viable sperm are present when the egg arrives, conception can occur.

54 What needs to happen before a fetus can be considered viable? **a. All organ systems must be functional.** This generally occurs by the third trimester, but even then a premature birth can have serious health consequences for the newborn.

55 Describe one new fetal development for each month:

3rd month: Bones begin to ossify, body growth accelerates while head growth slows, lungs begin to practice breathing amniotic fluid, external genitalia visible in male, 4 inches long and about 1 ounce

4th month: Body grows rapidly, legs lengthen, joints begin to form, face looks more human, roughly 7 inches long and 4 ounces

5th month: Mother can feel fetal movement, hair grows on head, lanugo covers skin, about 12 inches long and ½ to 1 pound

6th month: Eyebrows and eyelashes form, weight gain accelerates, roughly 11 to 14 inches long and just under 1½ pounds

7th month: Subcutaneous fat begins to form, eyelids open, usually turns to upside-down position, between 13 and 17 inches long and 2½ to 3 pounds

8th month: Subcutaneous fat increases, fetus appears more "babylike," testes of male descend into scrotum, 16 to 18 inches long and roughly 5 pounds

9th month: Substantial "plumping" with subcutaneous fat, lanugo shed, fingernails extend to fingertips, average newborn is 20 inches long and weighs 7½ pounds

56. Which hormones play an active role in the progression of labor? **d. Prostaglandins and Oxytocin.** Progesterone blocks uterine contractions. Its decline at the end of pregnancy allows the labor.

57. Which stage of labor usually lasts the longest? **a. stage 1, phase 1.** While certainly not the most painful, early labor can last upwards of an entire day.

58. Explain how childbirth is an example of a positive feedback loop. **In negative feedback, the body's response to a stimulus is to push in the opposite direction; for example, the body sweats in response to a temperature increase in order to bring it down. In positive feedback, the body's response is to push further in the same direction. There are numerous players in this loop but the bottom line is: Contractions get you more contractions which, in turn, get you more contractions.**

59. An individual's brain is considered fully developed **d. during young adulthood.** This doesn't mean you lose the capacity to learn, just that your brain is done developing the parts.

60. What is the formal term for a 20-month-old child? **a. Infant.** The terms "baby" and "toddler" aren't part of the formal medical lexicon, and a neonate is less than 4 weeks old.

61. Assume adult responsibilities, possibly including marriage and a family: **e. Young adult**

62. Faced with survival, it must process food, excrete waste, and obtain oxygen: **a. Neonate**

63. Primary and secondary sex characteristics begin to appear: **d. Adolescent**

64. Experiences senescence: **g. Old adult**

65. Deciduous teeth begin to form: **a. Infant**

66. Women go through menopause: **f. Middle-aged adult**

67. From 2 years of age to puberty: **c. Child**

6

The Part of Tens

Brush up on the ten best ways to study anatomy and physiology effectively.

Explore ten fun physiology facts in a free-wheeling testament to trivia.

Chapter **18**

Ten Study Tips for Anatomy and Physiology Students

What's the best way to tackle anatomy and physiology and come out successful on the other side? Of course, a good memory helps plenty — after all, you have to remember what goes where and which terminology attaches to which part. But with a little advance planning and tricks of the study trade, even students who complain that they can't remember their own names on exam day can summon the right terminology and information from their scrambled synaptic pathways. In this chapter, we cover ten key things you can start doing today to ensure success not only in anatomy and physiology but in any number of other classes.

Writing Down Important Stuff in Your Own Words

This is a simple idea that far too few students practice regularly. Don't stop at underlining and highlighting important material in your textbooks and study guides: Write it down. Combine your notes with the class notes. Whatever you do, don't just regurgitate it exactly as presented in the material you're studying. Find your own words. Create your own analogies. Tell your own tale of what happens to the bolus as it ventures into the digestive tract. Detail the course followed by a molecule of oxygen as it enters through the nose. Draw pictures of the differences between meiosis and mitosis. When you're answering practice questions, pay special attention to the ones you get wrong. Write reflections about why you answered incorrectly and what you need to remember about the right answer.

Unfortunately, typing the information doesn't create as strong of a connection so get that pencil, pen, marker, or even crayon moving! Completely relax into the process with the knowledge that no one else ever has to see what you write or sketch. All anyone else will ever see is your successful completion of the course!

Gaining Better Knowledge through Mnemonics

Studying anatomy and physiology involves remembering lists of terms, functions, and processes. Sprinkled throughout this book are suggestions to take just the first letter or two of each word from a list to create an acronym. Occasionally, we help you go one step beyond the acronym to a clever little thing called a *mnemonic device*. Simply put, the mnemonic is the thing you commit to memory as a means for remembering the more technical thing for which it stands. For example, a question in Chapter 7 asks you to list in order the epidermal layers from the dermis outward; we suggest that you commit the following phrase to memory: Be Super Greedy, Less Caring. Just like that, a complicated list like basale, spinosum, granulosum, lucidum, corneum gets a little closer to a permanent home in your brain.

EXAMPLE

Not feeling terribly clever at the moment you need a useful mnemonic? Surf on over to www. medicalmnemonics.com, which touts itself as the world's database of these useful tools. Here's a sampling of the site's offerings:

>> To remember the valves of the heart: "Try Pulling My Aorta: **T**ricuspid, **P**ulmonary, **M**itral, **A**ortic."

>> To remember the bones of the eye socket: "My Little Eye Sits in the orbit: **M**axilla, **L**acrimal, **E**thmoid, **S**phenoid."

>> To remember the upper arm muscles: "3 Bs bend the elbow: **B**iceps, **B**rachialis, **B**rachioradialis."

>> To remember the path out of the body from the top of the intestines: "Dow Jones Industrial Average Closing Stock Report: **D**uodenum, **J**ejunum, **I**leum, **A**ppendix, **C**olon, **S**igmoid, **R**ectum."

Discovering Your Learning Style

Every person has his or her own sense of style, and woe betide anyone who tries to shoehorn the masses into a single style. The same, of course, is true of students. To get the most out of your study time, you need to figure out what your learning style is and alter your study habits to accommodate it. No idea what we're talking about? Answer the questionnaire posted at www.vark-learn.com/english/page.asp?p=questionnaire, and the VARK guide to learning styles will tell you more about yourself than your last psychotherapist.

VARK, as you may have suspected, is an acronym. It stands for Visual (learning by seeing), Aural (learning by hearing), Reading/Writing (learning by reading and writing), and Kinesthetic (learning by touching, holding, or feeling). If you're a visual learner, you may get more out of anatomy and physiology by seeing the real thing in the flesh. If you're an aural learner, you may learn best in the classroom as the teacher lectures. If you're a reading and writing kind of learner, you'll get the most out of our first tip to write stuff down. And if you're a kinesthetic learner, there's nothing like touching or holding to commit something to memory.

TIP

Study in ways that match your learning style. Visual learners don't just benefit from pictures but also converting information into flowcharts. Aural learners need to say everything out loud when practicing rather than just write it down. Reading/Writing learners just read, write, and repeat. Kinesthetic learners need to make their studying as active as possible. Most K learners benefit from visual-based study techniques, particularly flow charting and grouping information. It can be difficult to touch concepts, but something as simple as kicking a ball around when reviewing can be helpful. This also holds true for those learners who tend to drift off to la-la land easily.

Getting a Grip on Greek and Latin

TIP

If you keep thinking "It's all Greek to me," congratulations on your insight! The truth of the matter is that most of it actually *is* Greek. So dust off your foreign language learning skills and begin with the basic vocabulary of medical terminology. Get started with the Greek and Latin roots, prefixes, and suffixes that appear on this book's Cheat Sheet at www.dummies.com/cheatsheet/anatomyphysiologywb. You'll soon discover that for every "little" word you learn, a whole mountain of additional terms and phrases are just waiting to be discovered.

Connecting with Concepts

It happens time and again in anatomy and physiology: One concept or connection mirrors another yet to be learned. But because you're focusing so hard on this week's lesson, you lose sight of the value in the previous month's lessons. For example, a concept like metabolism comes up in a variety of ways throughout your study of anatomy and physiology. When you encounter a repeat concept like that, create a special page or two for it at the back of your notebook or link the concept to a separate computer file. Then, every time the term comes up in class or in your textbook, add to the running list of notes on that concept. You'll have references to metabolism at each point it comes up *and* you'll be able to analyze its influences across different body systems.

Forming a Study Group

If you're really lucky, someone in your class (or maybe it's even you) has already suggested forming that time-honored tradition — a *study group*. The power of group members to fill gaps in your knowledge is priceless. But don't restrict it to late-night cramming just before each test. Meet with your group at least once a week to go over lecture notes and textbook readings. If it's true that people only retain about 10 percent of what they hear or read, then it makes sense that your fellow group members will recall things that slipped immediately from your mind.

Outlining What's to Come

As you read through a chapter of your textbook to prepare for the next lecture, prepare an outline of what you're reading, leaving plenty of space between subheadings. Then, during the lecture, take your notes within the outline you've already created. Piecing together an incomplete puzzle shows you where the key gaps in your knowledge may be. Additionally, familiarizing yourself with the upcoming vocabulary will lead to easier understanding when the concepts (especially those detailed processes) are presented.

Putting In Time to Practice

REMEMBER

Ten minutes a day goes a long way! Flashcards, mnemonic drills, practice tests — be creative and practice, practice, practice! Sometimes instructors share tidbits about what they plan to emphasize, but sometimes they don't. In the end, if you've done the work and put in the time to study and practice with information outside of class, the exact structure and content of an exam shouldn't make much difference. Nailing down the labeling and vocabulary early leaves you more time to work with the complicated physiology as you approach that all important test.

Flashcards are a highly effective strategy, but only if you do it right. Before starting, map out your timeline: When do you need to have all these terms down, and how often will you review them? For example, if you have one week, then you could review twice per day. During the first run-through, place the cards into two piles: correct and incorrect. The incorrect pile stays in the first stack and the correct pile moves up to the once-a-day stack. Later, pick up that first stack and repeat; you get to ignore the other one. For the next day, start with the once-daily pile and make the correct/incorrect piles like always. Correct ones get moved up another notch (every two days); incorrect ones goes back down to twice daily. Then go through the twice daily stack. This way, you don't waste valuable study time on terms you already have committed to memory.

Sleuthing Out Clues

Okay, it's test time! Take advantage of the test itself. You may find that the answer to an exam question that stumps you is revealed — at least partially — in the phrasing of a subsequent question. Make skipping questions part of your test-taking routine. (Do you even have one? Or do you just sit down and take the test?) Don't guess until you have to. That way, you will be more alert to these blessed little gifts even when you think that you already understand all the anatomical structures and physiological processes.

Reviewing Your Mistakes

The test is done and the grades are in. So there was a really tough question on the test and you blew it big-time? Don't beat yourself up over it. It's hardly a missed opportunity — this is where rolling with the punches really pays off. Go back over the entire test and pay extra attention to what you got wrong. Start your next practice sessions with those questions, and stay alert for upcoming material that may trip you up in a similar way. Remember that learning from a mistake leads to a "stronger" memory than having memorized correctly in the first place. You're not likely to make that mistake again come final exam time.

Chapter **19**

Ten Fun Physiology Facts

Basic anatomy may be fairly straightforward, but how the human body uses all those parts can present a smorgasbord of interesting discoveries. In this chapter, we give you just a peek at ten of the more intriguing aspects of what makes the body tick.

Boning Up on the Skeleton

Make no bones about it — the human skeleton is a trove of trivia. Try a few of these on for size (and check out Chapter 7 for the scoop on the skeleton):

» People are born with 300 bones in their infant bodies, but by the time they're adults they only have 206. Why? Because human bodies spend infancy and childhood putting the finishing touches on their skeletons, knitting together two or more bones into one. The only bone fully grown at birth is in the ear (the stapes). Plus, over a period of about seven years, each bone in the body is slowly replaced until it is a new bone.

» You're taller in the morning. No, really, you are. After you stand up from bed and get your day going, the cartilage between your back bones becomes compressed over time. By the end of the day, you're about one centimeter shorter.

>> Bones make up about 14 percent of an average individual's weight, yet they're as strong as granite. A block of bone the size of a matchbox can support 9 tons — four times more weight than concrete.

>> Bones can self-destruct. Excessive exposure to cadmium can prompt them to do so by triggering premature *apoptosis,* the controlled death of cells that takes place as part of normal growth and development. Alternatively, if you're not taking in enough calcium, certain hormones can leach calcium from your own bones to try to balance the blood's supply of this essential mineral.

Flexing Your Muscles

Besides your heart, the strongest muscle in the body (compared to its size) is your tongue. Hey, something has to counterbalance those 200 pounds of force the jaw muscles can deliver when you're chewing. And which muscles move more each day than even your heart? The ones surrounding your eyes, called the *extraocular muscles.* At three moves per second, they clock in more than 100,000 motions per day.

Every time you take a step, you employ around 200 of the roughly 600 muscles in your body. Researchers estimate that every minute you spend walking extends your life by anywhere from 90 seconds to two minutes. No wonder doctors advise that people shoot for 10,000 steps each day!

Thinking you'll just lift weights instead? Choose free weights over using a weight machine. The action of balancing the weights yourself rather than letting a machine do it for you builds muscle mass faster. (Chapter 8 introduces you to the muscles.)

Fighting Biological Invaders

As the body's internal defense network, the immune system sets blood and lymph against biological invaders seeking to colonize you. But its first line of defense is also the body's largest organ: the skin. If your skin — and some of the "friendly" bacteria living there — didn't secrete natural antibacterial compounds, you could wake up in the morning coated in microbial slime. As it is, though, most of the pathogens that land on you die quickly. (See Chapter 6 for details on the skin.)

The skin isn't the only thing working against the invading microbial hordes; the enzyme lysozyme found in human tears, saliva, and mucus is custom-made to disintegrate bacterial cell walls. (Remember: As eukaryotes, people's cells don't have walls.)

Not that all microorganisms are bad. You have between 2 and 5 pounds of bacteria living inside you, much of it in the intestines. As scientists have begun to understand what that microbial life is up to, it has become clear that your internal "microbiome" is a big part of what keeps you healthy. An imbalance in your gut flora could obviously lead to digestive issues but also vitamin deficiencies and may even influence mental health.

Cells Hair, There, and Everywhere

Every human alive today spent about 30 minutes at the start as a single cell. Now, however, your body is making 25 million new cells every day, and you shed and regrow all your outer skin cells about every 27 days. (See Chapter 6 for more about the skin.) Beards are the fastest-growing hairs on the human body; but all hairs have a genetically predetermined maximum length. The average human scalp has 100,000 hairs — blondes have more hair, on average, than dark-haired people do — and you lose between 40 and 100 strands of hair each day. Fingernails grow nearly four times faster than toenails do.

That's quite a bit of cellular production going on. Ever wonder what happens to those cells when they die? Well, they don't turn to dust. That's a commonly held belief that is actually just a myth. (Dust is mostly good, old-fashioned dirt.) While a small percentage could certainly be dead skin cells, we lose most of them when we bathe. Another common myth is that your hair and nails continue growing after death. Growth does appear to take place, but it's caused by the skin losing its bulk as the water evaporates. So the skin shrinks back, making it look like the hair or nails are growing out.

Swallowing Some Facts about Saliva and the Stomach

Besides being a digestive kick-starter, saliva prevents tooth decay and keeps your throat and mouth from drying out. It may not feel like it, but those six little glands make nearly half a gallon — about 1.5 liters — of saliva every day. Over the course of a lifetime, that's enough to fill about two average-sized swimming pools.

Hard to stomach? Consider this: With hydrochloric acid inside that's so corrosive it can dissolve wood and steel, your stomach should consume itself. But it doesn't. Why? Because you make your own natural supply of antacid. The epithelial cells lining your stomach secrete a steady supply of bicarbonate that neutralizes stomach acid on contact. Plus, these cells have a short life span (about three days) so the lining is continuously being replaced.

Be forewarned, though: That soothing bicarbonate produces carbon dioxide, among other things, as a by-product. That has to go somewhere. Some of it comes up in the form of a belch. And some of it contributes to the 17 ounces per day of flatulence that the average healthy human releases. (See Chapter 14 for an introduction to the digestive system.)

Appreciating the Extent of the Cardiovascular System

We tell you in Chapter 11 about the heart, the hardest-working muscle in the entire body, and its web of liquid connective tissue more commonly known as blood. But we don't mention a few facts that may make you appreciate your cardiovascular system even more.

In one day, the average individual's heart exerts enough power to lift a 1-ton weight more than 40 feet off the ground. During that day's pumping, the body's red blood cells travel about 12,000 miles, or roughly half the distance around the Earth at the equator and they complete a full lap every 60 seconds. They've got plenty of room to roam; if you laid every blood vessel in a body end-to-end, they would stretch around the Earth more than two times!

If that's hard to believe, consider this: Every square inch of skin contains 20 feet of blood vessels. Tissue the size of a pinhead contains 2,000 to 3,000 capillaries. You certainly have plenty of blood to fill all that space; you have 2.5 trillion red blood cells, more or less, in your body at any given moment, and your bone marrow creates 100 billion new ones every single day. That's so much blood that it would take 1.2 million mosquitoes each drinking their fill once to completely drain the average human of blood.

"You're Glowing" Isn't Just an Expression

You may have heard the expression before when you just got a raise, found out you're pregnant, or something awesome has happened and people say, "Look at you. You're glowing!" Turns out, that's not just an expression — you very well might be glowing. Scientists confirmed in 2009 that humans generate *bioluminescence*, just not at an intensity that our eyes can pick up on. Photons, or particles of light, are generated by excited molecules from cellular respiration. The glow is most prevalent in the late afternoon around the forehead, neck, and cheeks.

Another feature we can't see are our stripes. Called *Blaschko's lines*, these stripes are generated by the normal embryonic development of skin and are the same pattern for everyone. Most of us will never see them, but some skin conditions will follow the lines, making them more evident.

Looking at a Few of Your Extra Parts

With so many industrious components keeping you moving through your life, it can be startling to think about the number of body parts that, frankly, you just don't need. The first ones that come to mind often are the appendix and the wisdom teeth. The appendix, which does produce a few white blood cells and house important gut bacteria, generally gets to stay put so long as it doesn't get infected. But that third set of molars does little but cause pain and push around the teeth you really need.

You also have a "tailbone," or coccyx, which serves no function other than as a high-scoring Scrabble word, and it's a painful reminder of hard falls on your backside. It's a collection of fused vertebrae that researchers believe are what's left of the mammalian tail our evolutionary ancestors once sported. Other features that harken back to our early ancestors are the arrector pili that give us goosebumps (that don't actually keep us warm) and extrinsic ear muscles (though a small portion of people maintain control of these and can do a pretty neat ear wiggle).

People have a few other vestigial organs. Men, believe it or not, have an undeveloped vestigial uterus hanging on one side of the prostate gland, while women have a vestigial vas deferens — a cluster of tubules near the ovaries that would have become sperm ducts if they had inherited a Y chromosome from their fathers. (Part 5 has more details on both the male and female reproductive systems.)

Understanding Your Brain on Sleep

Superlatives about the human body tend to center around the brain. Fastest, neediest, most powerful — it is, after all, what makes us human. That 3-pound hunk of tissue demands 20 percent of the oxygen and calories the body takes in, communicates with 45 miles of nerves in the skin, contains individual neurons that can live more than a century, and generates enough energy when a person is awake — between 10 and 23 watts — to power a light bulb.

But the fun really begins when humans go to sleep. Scientists long have wondered why the body requires that people spend one-third of their lives unconscious. Although an individual may look quiet when she's with the sand man, studies have shown that the brain is more active in sleep than awake. Sleep, and the process of dreaming, helps the brain consolidate memories in ways that researchers are only beginning to understand.

For most of us, communication with the cerebellum is shut off by the pons so that we don't physically make the movements we are making in our dreams. But if you know anyone with *parasomnia* (sleep walking, sleep talking, and so on), you know this switch doesn't always work — likely because it's way more complicated than a single "switch."

Getting Sensational News

When people talk about sensation, most of the time they're referring to the five primary senses — vision, hearing, taste, touch, and smell (see Chapter 9). But the notion of a "sixth sense" may be more than the stuff of science fiction. Vision combines senses for both light and color, and there is growing evidence that your *proprioception* — the ability to detect your relative position in space — may rely on your body's ability to detect magnetic fields in much the same way migrating birds do. Blind people have been known to develop echolocation, or the ability to hear subtle changes in sound bouncing back from otherwise unseen objects. And stress sometimes causes people to experience time dilation.

In fact, your senses are more subjective than you may like to admit. Things get even more complex when you consider the condition known as *synesthesia*, in which a person can "hear" color or "see" sound. The eyes, which can distinguish up to one million colors, routinely take in more information than the largest telescope ever created. The nose can identify and the brain subsequently can remember more than 50,000 smells. Static touch can discern an object about twice the diameter of an eyelash, while dynamic touch — dragging a finger along a surface — can detect bumps the size of a very large molecule.

Index

frontal bone, 106
frontal lobe, 172
funiculi, 169

G

gallbladder, 279
gamete production, 319–320
ganglion, 161
gap junctions, 64, 218
gastric juice, 275
gastrula, 266
gels, 43
gene expression, 35
gene mutation, 59
general adaptation syndrome, 202
general sense receptors, 181
gingivae (gums), 269
glenoid fossa, 113
glial cells (neuroglia), 73, 158, 161
gliding (plane) joints, 118
glomerular filtration, 292
glycolysis, 28
goblet cells, 251, 278
golgi apparatus (golgi body), 44
gomphosis joints, 117
gonads, 194
goose bumps (goose pimples), 87
gray commissure, 169
gray matter, 169
greater omentum, 278
ground substance, 68
growth hormone (GH), 103, 195, 330
growth plate, 98
gyri, 171

H

hair, 87, 89–91, 94, 347
hair cells, 183
hair plexus, 86
Haversian canal (central canal), 69, 99

hematopoiesis (hemopoiesis), 96
hemoglobin, 69–70, 81
hilum, 235
hinge joints, 118
histology
    cell junctions, 63–64
    connective tissue, 68–71, 74
    defined, 63
    epithelial tissue, 64–67, 74
    muscle tissue, 71–72, 75
    nervous tissue, 73, 75
histones, 55
homeostasis, 179, 202–203, 206
homeostatic imbalance, 202
hormones, 191–193
humerus, 113
humoral immunity, 243
hyaline cartilage, 69
hydrogen bonds, 20–21
hydrolysis, 21
hydrophilic polar heads, 37
hydrophobic nonpolar tails, 37
hydrostatic pressure, 38
hyoid bone (tongue bone), 105
hyperpolarization, 166
hypertension, 225
hypertonic solutions, 39
hypodermis (superficial fascia; subcutaneous tissue), 80
hyponychium, 88
hypophyseal tract, 195
hypotension, 225
hypothalamus, 171, 194, 254
hypotonic solutions, 39

I

ileum, 278
ilium, 113
immunity, 242–244, 246–247, 346
impulses, 136

incus, 106, 183
inferior nasal concha bones, 106
inferior vena cava, 210, 213
infundibulum, 194
inhibitory neurotransmitters, 167
innate (non-specific) defenses, 242
inner tunic, 182
insertion, 128
integumentary system
  dermis, 82–85, 93–94
  epidermis, 80–85, 93–94
  glands, 88–89, 91–92, 94
  hair, 87, 89–91, 94
  nails, 88, 90, 92, 94
  overview, 79–80
  receptors, 86, 94
interatrial septum, 213
intercalated discs, 71, 131, 218
interferons, 242
interleukin-2, 243
intermediate filaments, 43
internal ear, 183
interoreceptors, 181
interphase, 54
interstitial (extracellular) fluid, 63, 225, 231–232
interventricular septum, 214
intervertebral foramina, 108
intestinal lipase, 278
intracellular matrix, 69
intramembranous ossification, 103
invaginations, 239
inverting enzymes, 278
involution, 239
ionic bonds, 19
ions, 17
irises, 181
irregular bones, 98
ischium, 113
islets of Langerhans, 194, 199, 280

isometric contraction, 140
isotonic contraction, 140–141
isotonic solutions, 38
isotopes, 16
isthmus, 198

J

jejunum, 278
joint capsules, 118
joints, 117–121, 125
jugular foramen, 106

K

keratin, 43, 80
keratinocytes, 81
keratohyalin, 81
keto acids, 29
kidneys, 288–291, 299
Krebs cycle (citric acid cycle), 29–30
Kupffer cells, 279

L

labia majora and minora, 318
labial frenula, 269
labial glands, 269
lacrimal bones, 106
lacrimal glands, 255
lacteals, 232
lactic acid, 30, 143
lacunae, 69, 99
lambdoidal suture, 106
lamellae, 69, 99
lamellated (Pacinian) corpuscles, 80
laminae, 107
Langerhans cells, 81
large intestine (colon), 282–283
large motor units, 136
larynx, 257–258
leukocytes, 69–70, 96

mixed spinal nerve, 176
mnemonic devices, 340
molecules, 19
mons pubis, 318
morphogenesis, 128
motor (efferent) system, 158
motor end plate, 136
motor neurons, 129
mouth, 268–274, 284–285
mucosa, 266
multinucleated muscle fibers, 71
multipolar neurons, 159
multipotent stem cells, 35
muscle fatigue, 30
muscle spindles, 131
muscle tissue, 71–72, 75
muscle twitch, 140
muscles
  anatomy of, 133–138, 150–151
  characteristics of, 127
  classifications of, 131–133, 150
  fun facts, 346
  functions of, 128
  lever action, 142–144, 152
  muscle tone and function, 140–142, 151–152
  names of, 144–149, 152–153
  as organs, 138–140, 151
  overview, 127–130, 150
  terminology, 128
muscularis, 266
myelin, 166
myelin sheath, 160
myocardium, 211
myocytes, 71
myofibrils, 71, 129
myogenesis, 128
myoglobin, 143
myology, 128
myosin, 71, 133–135

N

nails, 88, 90, 92, 94
nasal bones, 106
nasal cavity, 255
nasopharynx, 274
natural killer cells (NKs), 242
nephrons, 292–293
nerve fibers, 160
nerve impulses, 158, 165–166
nerve tracts, 169
nerves, 158, 160
nervous system
  autonomic nervous system, 178–180, 189
  central nervous system, 168–175, 188–189
  communication process, 164–168, 187–188
  components of, 158–164, 187
  divisions of, 158
  overview, 157–158
  peripheral nervous system, 176–178, 189
  senses, 181–186, 189–190
nervous tissue, 73, 75
neurofibrillae, 159
neurolemma (neurilemma), 160
neurons, 73, 158–160, 165–166
neurotransmitters, 159, 164–167
neutralization, 243
neutrons, 16
Nissl bodies, 158–159
nociception, 86
nodes of Ranvier, 166
norepinephrine (noradrenaline), 197–198
nose, 254–257, 263
nostrils (external nares), 254
nuclear envelope, 41
nuclear lamina, 41
nucleic acids, 23
nucleolus, 41
nucleoplasm, 41
nucleus, 16, 35, 41–42, 50–51

O

oblique muscle, 266
occipital bone, 106
occipital lobe, 172
olecranon, 113
olfactory bulb, 254
olfactory cortex, 254
olfactory foramina, 106
oligodendrocytes, 161
online resources
  Cheat Sheet (companion to book), 3
  mnemonic devices, 340
  online practice (companion to book), 3
  VARK guide to learning styles, 341
optic chiasma, 171
optic disc, 182
optic foramen, 106
oral cavity, 269–271
organelles, 35, 43–47, 51
organic compounds, 21–23
oropharynx, 270
os coxae, 113
osmolarity, 38
osmoreceptors, 195
osmosis, 38–39
osmotic pressure, 38
osseus (bone) tissue, 69
ossification, 103–105, 123–124
osteoblasts, 69, 103
osteoclasts, 103
osteocytes, 69, 99
osteology, 95
osteons, 99
ovaries, 318–320
ovulation, 319
oxaloacetic acid (OAA), 29
oxidative phosphorylation, 29
oxygenation, 210

oxygen-carbon dioxide exchange, 260–262, 264
oxytocin (OT), 195, 329

P

pacinian corpuscles, 86
palate, 270
palatine bones, 106
palatine tonsils, 239, 270
palpebrae (eyelids), 182
pancreas, 194, 199, 280
Paneth cells, 278
papillary layer, 82
papillary muscles, 214
paracrine factors, 192
paracrine glands, 192
parafollicular cells (C cells), 198
paranasal sinuses, 107, 255
parasympathetic nervous system, 178–179
parathyroid, 194
parathyroid hormone (PTH), 104, 198
parietal bones, 106
parietal lobe, 172
parietal membrane, 8
parietal pericardium, 211
parietal peritoneum, 8
parietal pleura, 251
parturition (childbirth), 329
passive transport, 38
patellae, 113
patellar reflex, 177
pectinate muscles, 213
pectoral girdle, 113
pedicles, 107
pelvic girdle, 113
penis, 303–306
peptidase, 278
peptide bonds, 47

thyrotropic hormone (thyroid-stimulating hormone [TSH]), 195

thyroxine, 198

tibia, 113

tight junctions, 63

tone (tonus), 129, 140

tongue, 271

totipotent stem cells, 35

trabeculae, 99, 235

trabeculae carneae, 214

trabecular (spongy) bone, 98–99

trachea, 251, 274

transfer RNA (tRNA), 47

transitional epithelium, 66

transport vesicles, 44

transverse foramina, 108

transverse processes, 107

tricuspid valve, 214

trigeminal nerve, 107

trigger zone, 165

triiodothyronine, 198

tropic hormones, 195

tropomyosin, 135

troponin, 135

true ribs, 108

trypsin, 278

trypsinogen, 278

t-t complex, 135

t-tubules, 136

tubulin, 43

tympanic membrane (eardrum), 65, 182–183

U

ulna, 113

unipolar neurons, 159

ureters, 296

urethra, 296–297

urinary system

  filtration, 292–295, 299–300

  kidneys, 288–291, 299

  overview, 287–288

  urination, 296–298, 300

uterine tubes (Fallopian tubes; oviducts), 318

uterus, 318, 320–321

V

vagina, 318

vasoconstriction, 225

vasopressin (antidiuretic hormone [ADH]), 195, 294

veins, 223

ventral root, 176

ventricles, 172, 214

venules, 223

vertebral foramen, 107–108

vesicles, 39

vestibule, 183, 254

virilism, 198

viscera, 72

visceral membrane, 7

visceral pericardium, 8, 211

visceral peritoneum, 266

visceral pleura, 8, 251

vitreous humor, 181–182

Volkmann's canals, 99

voltage-gated sodium channels, 165

vomer, 106

W

Wernicke's area, 172

white matter (funiculus), 169

white pulp, 238

X

xiphoid process, 108

Z

zygomatic bones, 106

About the Authors

Erin Odya is an anatomy & physiology teacher at Carmel High School in Carmel, Indiana, which is widely regarded as one of Indiana's top schools. She takes great pride in her role as an educator. Erin is also the author of *Anatomy & Physiology for Dummies*, 3rd Edition.

Pat DuPree taught anatomy/physiology, biology, medical terminology, and environmental science for 24 years at several colleges and universities in Los Angeles County. She holds two undergraduate life science degrees and a master's degree from Auburn University and conducted cancer research at Southern Research Institute in Birmingham, Alabama, before joining the Muscogee Health Department in Columbus, Georgia. In 1970, she moved to Redondo Beach, California, where she was a university instructor and raised her two sons, Dave Dupree and Mark DuPree. Now Pat is retired and lives on lovely Pine Lake in rural Georgia with her husband of 55 years, Dr. James E. DuPree.

Author's Acknowledgments

Many thanks to my family and friends for their support, especially Dr. Sarah Brookes at Purdue University and Dr. Jane Black at Monash University. —Erin Odya

Dedication

To Gabe and his never-ending questions about how his 8-year-old body works and to all of my students — past, present, and future — who challenge my knowledge and inspire my curiosity. —Erin Odya

Publisher's Acknowledgments

Executive Editor: Lindsay Sandman Lefevere

Project Editor: Tim Gallan

Technical Reviewer: Charles Jacobs

Production Editor: Siddique Shaik

Cover Image: ©yodiyim/Getty Images

Take dummies with you everywhere you go!

Whether you are excited about e-books, want more from the web, must have your mobile apps, or are swept up in social media, dummies makes everything easier.

Find us online!

Leverage the power

Dummies is the global leader in the reference category and one of the most trusted and highly regarded brands in the world. No longer just focused on books, customers now have access to the dummies content they need in the format they want. Together we'll craft a solution that engages your customers, stands out from the competition, and helps you meet your goals.

Advertising & Sponsorships

Connect with an engaged audience on a powerful multimedia site, and position your message alongside expert how-to content. Dummies.com is a one-stop shop for free, online information and know-how curated by a team of experts.

- Targeted ads
- Video
- Email Marketing

- Microsites
- Sweepstakes sponsorship

20 MILLION PAGE VIEWS EVERY SINGLE MONTH

15 MILLION UNIQUE VISITORS PER MONTH

43% OF ALL VISITORS ACCESS THE SITE VIA THEIR MOBILE DEVICES

700,000 NEWSLETTER SUBSCRIPTIONS TO THE INBOXES OF

300,000 UNIQUE INDIVIDUALS EVERY WEEK

# of dummies

Custom Publishing

Reach a global audience in any language by creating a solution that will differentiate you from competitors, amplify your message, and encourage customers to make a buying decision.

- Apps
- Books
- eBooks
- Video
- Audio
- Webinars

Brand Licensing & Content

Leverage the strength of the world's most popular reference brand to reach new audiences and channels of distribution.

For more information, visit dummies.com/biz

 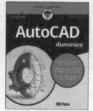     

Learning Made Easy

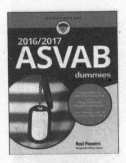